T0205572

Integration of Air Conditioning and Heating into Modern Power Systems

Yi Ding · Yonghua Song ·
Hongxun Hui · Changzheng Shao

Integration of Air Conditioning and Heating into Modern Power Systems

Enabling Demand Response and Energy Efficiency

Yi Ding
Zhejiang University
Hangzhou, Zhejiang, China

Hongxun Hui
Zhejiang University
Hangzhou, Zhejiang, China

Yonghua Song
University of Macau
Macau, Macao

Zhejiang University
Hangzhou, Zhejiang, China

Changzheng Shao
Zhejiang University
Hangzhou, Zhejiang, China

ISBN 978-981-13-6422-8 ISBN 978-981-13-6420-4 (eBook)
https://doi.org/10.1007/978-981-13-6420-4

Library of Congress Control Number: 2019930966

This Springer imprint is published by the registered company Springer Nature Singapore Pte Ltd.
The registered company address is: 152 Beach Road, #21-01/04 Gateway East, Singapore 189721, Singapore

Preface

The release of carbon dioxide and other greenhouse gases due to human activity results in a host of environmental issues. The enhanced public concern for adverse environmental impacts associated with the use of conventional energy sources requires a transition toward clean energy systems. Moreover, the de-carbonization of electric power systems plays a significant role in reducing anthropogenic carbon emissions, since electric power systems remain the primary source of carbon emissions in the world. As a result, the application of renewable energy in electric power systems generates great interest. Among renewable energy sources, wind energy has experienced rapid development and has made significant inroads into electrical power systems. Over the past decade, the global cumulative installed capacity of wind energy has been growing at a rate of more than 21% annually. In 2015, global wind power capacity increased by about 17%. In China, wind power has become the third largest power source, following thermal and hydroelectric power, and generates 4.8% of the country's electricity in 2017.

However, the power generated from renewable energy such as wind is fluctuating and uncertain, which presents significant challenges to the efficient utilization. As electricity demand and supply must be maintained in balance at all times, power systems need to absorb the electricity fluctuation from renewable energy. An increasing capacity of fluctuating renewable energy will increase the need for flexibility during power system operation. Flexibility is the ability of the power system to deploy its resources for rebalancing customer demand and generation when fluctuations exist. For example, downward reserve is required to ensure power system balance when the amount of injected wind power is higher. Conversely, upward reserve is required when the amount of wind power injection is lower. If there is not sufficient operational flexibility, the efficient utilization of renewable energy cannot be achieved. The serious wind power curtailment issue in China could well prove that. The coal-dominated generation mix in China works against the high level of wind penetration, since the flexibility of coal-fired generating units is constrained by their ramp-up and ramp-down rates as well as their minimum stable generation output. China's inflexible generation mix, which cannot respond well to changes in wind power output, forces it to curtail a large amount of wind energy every year, despite the country's renewable energy ambitions. Wind energy curtailment in China is becoming increasingly serious. The total energy loss from wind curtailment from 2011 to 2015 was approximately 95.9 billion kWh, nearly equal to the gross electricity generated by wind energy in Denmark in 2013.

The development of information and communication technologies and electricity market has made the remote control of flexible loads much easier. Thus, it is possible for small end customers to provide operating reserve to support power system operation. As one of the most popular and easily controlled flexible loads, air conditioners and heating equipment account for a large share in power consumption due to the mass application across the world. Statistical data have shown that air conditioners account for approximately 35, 33, and 40% of the electricity consumption during the peak hours in many cities in China, Spain, and India, respectively. Therefore, these flexible loads have yielded enormous potential in serving as energy storage devices, which can provide operating reserve by reducing power consumption temporarily. In this field, some researches have been conducted.

The book focuses on integration of air conditioning and heating as demand response into modern power system operation and planning. Both models and methods have been addressed with engineering practice. This is achieved by providing in-depth study on air conditioner aggregation providing operating reserve and frequency regulation service for helping power system operation. Different models of air conditioner aggregation and corresponding control methods are studied in detail. Moreover, the comprehensive and systematic treatment of incorporating flexible heating demand into the integrated energy system is one of the major features of the book, which is particularly suited for readers who are interested to learn methods and solutions of demand response in smart grid environment. The book can benefit researchers, engineers, and graduate students in the fields of electrical and electronic engineering, control engineering, computer engineering, etc.

There are eight chapters in this book.

- Chapter 1 introduces the development of the air conditioning and heating loads as demand response in modern power systems. The advantages and some existing studies are also introduced in this chapter.
- Chapter 2 is devoted to the operating reserve evaluation of aggregated air conditioners. The thermal model of the room and the operating reserve characteristics of an individual air conditioner are developed. The performance of the operating reserve provided by aggregated air conditioners and the corresponding evaluation indexes are proposed, respectively. The numerical studies are presented to illustrate the effectiveness of the proposed model and methods.
- Chapter 3 is devoted to the operating reserve capacity evaluation of aggregated heterogeneous thermostatically controlled loads (TCLs) with price signals. The individual TCL model on account of consumer behaviors is developed. On this basis, the moment estimation method and the probability density estimation method are proposed to estimate the reserve capacity with insufficient data.

- Chapter 4 is devoted to the lead–lag rebound effect from the aggregate response of air conditioners controlled by the changes of the set point temperature. The impacts of the lead–lag rebound are quantified by a proposed capacity–time evaluation framework of operating reserve. On this basis, an optimal sequential dispatch strategy of air conditioners is proposed for the entire mitigation of the lead–lag rebound and the provision of operating reserve with multiple duration time.
- Chapter 5 is devoted to the frequency regulation service (FRS) provided by the inverter air conditioners. The equivalent modeling of the inverter air conditioners is developed. In this manner, the inverter AC can be scheduled and compatible with the existing control system. A stochastic allocation method of the regulation sequence among inverter ACs is proposed to reduce the effect of FRS on customers. Besides, a hybrid control strategy by considering the dead band control and the hysteresis control is developed to reduce the frequency fluctuations of power systems.
- Chapter 6 expands the demand response to the heat and power integrated energy system (HE-IES). HE-IES, based on combined heat and power (CHP), is one of the most important forms of IES. It is assumed that both electricity energy system and heat energy system are managed by a single ISO and all the aggregators seek to minimize their energy costs. Incorporating the aggregators' flexible energy demand into the central energy dispatch model therefore forms a two-level optimization problem (TLOP), where the ISO maximizes social welfare subject to aggregators' strategies, in which aggregators adjust their energy demand so as to minimize the energy purchase cost.
- Chapter 7 analyzes the demand response potential of customers (usually refer to buildings) in the distribution-level heat and electricity integrated energy system. This chapter proposes a framework for utilizing the demand response to improve the operation of the integrated energy system which has gained rapid development recently. The framework involves three levels of the integrated energy system: aggregation of the smart buildings, distribution system, and transmission system or sub-transmission system. In the framework, the buildings' demand response potential can be fully utilized and the operational flexibility of the transmission-level integrated energy system can be significantly improved.
- Chapter 8 is devoted to the evaluation of the economy of the three different flexibility resources to find the advantages/disadvantages of different resources and to provide guidance for investment in these flexible resources.

The authors shared the work in writing this book.

It was a pleasure working with Springer Associate Editor, Ms. Jasmine Dou.

Hangzhou, China — Yi Ding
Macau/Hangzhou, China — Yonghua Song
Hangzhou, China — Hongxun Hui
Hangzhou, China — Changzheng Shao

Contents

1 Air Conditioning and Heating as Demand Response in Modern Power Systems . . . 1
References . . . 4

2 Aggregated Air Conditioners for Providing Operating Reserve . . . 7
2.1 Introduction . . . 7
2.2 Operating Reserve Provided by Individual AC . . . 9
2.2.1 Thermal Model of the Room . . . 9
2.2.2 Operation Characteristics of Individual Air Conditioner . . . 10
2.2.3 Operating Reserve Provided by Individual Air Conditioner . . . 11
2.3 Operating Reserve Provided by Aggregated ACs . . . 14
2.3.1 Performance of Operating Reserve Provided by Aggregated ACs . . . 15
2.3.2 Simulation Framework for Evaluating Operating Reserve Performance . . . 16
2.4 Case Studies and Discussions . . . 18
2.4.1 Parameter Initialization . . . 18
2.4.2 Operating Reserve Performance with Different Temperature Adjustments . . . 19
2.4.3 Operating Reserve Performance with Different Numbers of ACs . . . 21
2.4.4 Analysis of Aggregated ACs Returning to Original Set Temperature . . . 24
2.4.5 Analysis of Demand Response in Actual Case Studies . . . 25
2.5 Conclusions . . . 26
References . . . 26

3 Heterogeneous Air Conditioner Aggregation for Providing Operating Reserve Considering Price Signals . . . 29
3.1 Introduction . . . 29
3.2 Individual TCL Model . . . 30
3.2.1 Framework and Electric-Thermal Model . . . 30
3.2.2 Consumer Satisfaction Quantization . . . 32
3.2.3 Maximum Satisfaction Control Strategies . . . 33
3.3 ORC Evaluation of Aggregated Heterogeneous TCLs . . . 34
3.3.1 Moment Estimation Method . . . 35
3.3.2 Probability Density Estimation Method . . . 35
3.4 Case Studies . . . 38
3.4.1 The Test System . . . 39
3.4.2 ORC Evaluation with Insufficient Data . . . 41
3.4.3 ORC Evaluation in Actual Case Studies . . . 43
3.5 Conclusions . . . 45
References . . . 46

4 Air Conditioner Aggregation for Providing Operating Reserve Considering Lead-Lag Rebound Effect . . . 49
4.1 Introduction . . . 49
4.2 Analysis of the Lead-Lag Rebound Effect . . . 52
4.2.1 Model of an Individual AC . . . 52
4.2.2 Aggregate Response of ACs . . . 52
4.2.3 Lead Rebound Effect and Lag Rebound Effect . . . 54
4.3 Capacity-Time Evaluation of the Operating Reserve Considering Lead-Lag Rebound Effect . . . 56
4.3.1 Universal Expression of the Load Reduction/Increase . . . 56
4.3.2 Evaluation of the Operating Reserve Provided by ACs on the Capacity Dimension . . . 56
4.3.3 Evaluation of the Operating Reserve Provided by ACs on the Time Dimension . . . 58
4.4 Sequential Dispatch Strategy of ACs for Providing Operating Reserve with Multiple Duration Time . . . 58
4.4.1 The Interactions Among the System Operator, Aggregators and Consumers . . . 58
4.4.2 Sequential Dispatch Strategy of ACs to Mitigate the Lead-Lag Rebound Effect . . . 60
4.4.3 Capacity-Time Co-optimization of Sequential Dispatch Process During the Reserve Deployment Period . . . 62
4.4.4 Capacity-Time Co-optimization of Sequential Dispatch Process During the Recovery Period . . . 64

4.5 Case Studies and Simulation Results 66
4.5.1 Evaluation of ACs' Potential for the Provision of Operating Reserve 66
4.5.2 Provision of Operating Reserve with Various Duration Time and Reserve Capacity 70
4.5.3 Comparison of Different Dispatch Strategy of ACs for the Provision of Operating Reserve 74
4.6 Conclusions 79
References 79

5 Inverter Air Conditioner Aggregation for Providing Frequency Regulation Service 83
5.1 Introduction 83
5.2 Thermal and Electrical Model of the Inverter AC Considering Providing FRS 85
5.2.1 Thermal Model of a Room. 85
5.2.2 Electrical Model of an Inverter AC Considering Providing FRS for Power Systems 86
5.2.3 Analysis of the Thermal and Electrical Model 87
5.3 Equivalent Modeling of Inverter ACs for Providing Frequency Regulation Service 89
5.3.1 Equivalent Modeling of Inverter ACs 89
5.3.2 Equivalent Control Parameters 91
5.3.3 Equivalent Evaluation Parameters 92
5.4 Control of Aggregated Inverter ACs for Providing Frequency Regulation Service 93
5.4.1 The Regulation Capacity Allocation Among Generators and Inverter ACs 93
5.4.2 The Control Strategy of Inverter ACs 95
5.4.3 The Communication and Control Process of Inverter ACs 95
5.5 Case Studies. 98
5.5.1 Test System 98
5.5.2 Simulation Results. 99
5.5.3 Experimental Results 103
5.6 Conclusions 104
References 105

6 Integration of Flexible Heating Demand into the Integrated Energy System 107
6.1 Introduction 107
6.2 Heat and Electricity Integrated Energy System 109
6.2.1 Description of the HE-IES 109
6.2.2 Modelling the Customer Aggregators' Energy Demand . . . 110

6.3 TLOP-Formulation of the Dispatch Model 113
6.4 Simplifying the Sub-problems' KKT Conditions 115
6.5 Application and Test Results 119
6.5.1 Test System and Scenarios 119
6.5.2 Simulation Results 122
6.6 Conclusions 126
References 126

7 A Three-Level Framwork for Utilizing the Demand Response to Improve the Operation of the Integrated Energy Systems 129
7.1 Introduction 129
7.2 Energy Demand of Smart Buildings 131
7.2.1 Modeling Individual Building's Energy Demand Based on the Comprehensive DR Strategy 132
7.2.2 Energy Demand Aggregation of Multiple Buildings 133
7.3 Concept and Framework of the Real-Time DRX Market 134
7.3.1 Three-Level Framework of the DRX Market 135
7.3.2 Optimization Models in the DRX Market 137
7.3.3 Clearing of the DRX Market 140
7.4 Case Studies 143
7.4.1 Test System and Parameters 143
7.4.2 Comparison Between the Proposed Comprehensive DR Strategies and the Load Shifting Strategy 144
7.4.3 Comparison Between the Real-Time DRX Framework and the Day-Ahead DR Framework 148
7.4.4 Comparisons Between the Proposed Clearing Method and the Iteration-Based Clearing Method 149
7.5 Conclusions 150
References 150

8 Economical Evaluation of the Flexible Resources for Providing the Operational Flexibility in the Power System 153
8.1 Introduction 153
8.2 Methods to Calculate the Balancing Costs When Utilizing Different Flexible Resources 156
8.2.1 Mathematic Model for Evaluating the Balancing Cost of Utilizing Coal-Fired Generating Units 156
8.2.2 Optimization Model for Sizing the ESS and Determining the Balancing Cost 160
8.2.3 Optimization Model for Determining the Balancing Cost When Utilizing the DSM 162

8.3 Simulation Results and Analysis 164
8.3.1 Parameters 164
8.3.2 Simulation Results 165
8.3.3 Comparison and Conclusion 171
8.4 Conclusions 172
References 172

Abbreviations

AC	Air conditioner
AT	Activation time
CER	Carbon emission rate
CO_2	Carbon dioxide
COP	Coefficient of performance
DR	Demand response
DSR	Demand-side resource
DSM	Demand-side management
DT	Duration time
ECR	ESS's energy capacity requirement
EER	Energy efficiency ratio
ED	Economic dispatch
ESS	Energy storage system
FRS	Frequency regulation service
GHG	Greenhouse gas
HVAC	Heating, ventilating and air conditioning
ICT	Information and communication technologies
LCE	Expected load curtailment energy
LCR	Load power capacity requirement
ME	Moment estimation
MWhw	Per MWh of wind energy production
MWw	Per MW of installed wind power capacity
ORC	Operating reserve capacity
PCR	ESS's power capacity requirement
PDE	Probability density estimation
PER	Period
PI	Proportional integral
PFR	Primary frequency regulation
RES	Renewable energy source
RC	Reserve capacity

RT	Response time
RR	Ramp rate
SFR	Secondary frequency regulation
SCCR	System coal consumption rate
TCL	Thermostatically controlled load
TSK	Takagi–Sugeno–Kang
UC	Unit commitment

Symbols

H_{gain}	Total heat gains of the room
H_{loss}	Total heat losses of the room
H_{AC}	Heating/cooling generated by AC
$H_{internal}$	Heat gain from appliances and occupants
H_{solar}	Solar radiation received by the room
c_A	Heat capacity of air
ρ_A	Density of air
T_A	Temperature in the room
T_O	Ambient temperature
$T_{set}^{(k)}$	Set temperature of the *k*th AC
$T_{set2}^{(k)}$	Reset temperature of the *k*th AC
ΔT_{set}	Temperature adjustment
$T_{hy}^{(k)}$	Hysteresis band control of temperature
A	Living area of the room
h	Height of the room
A_S	Surface area of the room envelope
V	Volume of the room
ε	Coefficient of heat release by appliances and occupants
$P_{incident}$	Time-varying coefficient of heat absorbed from the sun
S	Operation state of AC
S_{cool}	Cooling state of AC
S_{heat}	Heating state of AC
S_{styb}	Standby state of AC
P	Power of AC
$P_{cool}^{(k)}$	Power of the *k*th AC in cooling state
$P_{styb}^{(k)}$	Power of the *k*th AC in standby state
$PD_{cool}^{(k)}$	Length of time in cooling state
$PD_{styb}^{(k)}$	Length of time in standby state

k	Serial number of ACs
N	Total number of ACs
t	Time
t_{ds}	Start moment of duration time
t_{de}	End moment of duration time
α	Valid interval of providing operating reserve
P_{max}	Maximum power before receiving the control signal
P_{min}	Minimum power after receiving the control signal
SS	Send control signal
SR	Receive control signal
t_{SS}	Moment of sending control signal
t_{SR}	Moment of receiving control signal
T_{set}	Set temperature of the TCL
T_{room}	Room temperature
S	Operation state of the TCL
ΔT	Hysteresis band of the room temperature
$P(t)$	Load power of the TCL
P_r	Rated power of the TCL
$p(t)$	Electricity price
$C(t)$	Electricity cost
$Q(t)$	Refrigerating capacity of the TCL
H_r	Equivalent thermal conductance of the room
T_{ext}	Outdoor ambient temperature
C_{room}	Thermal mass of the room
μ	Membership values
P^k_{avg}	Average power of the kth TCL
ΔP^k	Power deviation of the kth TCL
P_{ORC}	Total ORC provided by aggregated heterogeneous TCLs
$\hat{f}_h$	Dimension probability density estimation
N_s	TCL's number which can be measured
N	Total number of TCLs
$\widehat{P}_{avg}$	Evaluation value of the average power P_{avg}
$\widehat{P}_{ORC}$	Evaluation value of the total ORC P_{ORC}
α	Weight of consumer preferences between the electricity cost and the room temperature
e_m	Error of the estimated ORC
i	Index of an individual AC
g	Index of an AC group
k	Index of an AC group to be dispatched
q	Index of an AC group to be recovered
d	Index of the reserve deployment period
r	Index of the recovery period
t	Index of time (h)

$\theta_i(t)$	Room temperature corresponding to the ith AC at the time t (°C)
$\theta_a(t)$	Ambient temperature (°C)
$T_{set,i}$	Set point temperature of the ith AC (°C)
θ_i^+/θ_i^-	The upper/lower temperature hysteresis band of the ith AC (°C)
$\mathbf{\Gamma}$	The set of all the ACs under an aggregator
$N^{\max}$	The number of ACs in $\mathbf{\Gamma}$
τ_g^d/τ_g^r	The reserve deployment/recovery time instant of group g (h)
RC_g^d	Reserve capacity of group g during the reserve deployment period (MW)
RC_g^r	Reserve capacity of group g during the reserve recovery period (MW)
BC_g^d	Rebound capacity of group g during the reserve deployment period (MW)
BC_g^r	Rebound capacity of group g during the reserve recovery period (MW)
PD_g	The difference of the aggregate power before and after the changes of the set point temperature (MW)
$\underset{t_1 \to t_2}{SD}(x(t))$	The standard deviation of the variable $x(t)$ during the time period $[t_1, t_2]$
PV_g	Power volatility of aggregate power of group g after the end of rebound process (MW)
DT_g	Deployment duration of operating reserve provided by ACs in group g (h)
RT_g	Ramp time of operating reserve provided by ACs in group g (min)
RR_g	Ramp rate of operating reserve provided by ACs in group g (MW/min)
RC_g^*	Required reserve capacity instructed by the system operator (MW)
DT_g^*	Required deployment duration instructed by the system operator (h)
Q_{gain}	Total heat gains of the room (kW)
Q_{AC}	Refrigerating capacity of the AC (kW)
Q_{dis}	Heat power from people, lights, appliances, and other disturbances (kW)
P_{AC}	Operating power of the inverter AC (kW)
f_{AC}	Operating frequency of the inverter AC (Hz)
T_O	Ambient temperature
T_A	Indoor temperature
T_{set}	Set temperature of the inverter AC (°C)
T_{dev}	The deviation between the indoor temperature and the set temperature (°C)
T_c	Time constant of inverter AC's compressor (s)
ΔP_G	Regulation capacity of the generator (kW)
ΔP_{AC}	Regulation capacity of the inverter AC (kW)
Δf	The frequency deviation of the system (Hz)

β_i	The allocation coefficient of the regulation capacity among inverter ACs
R	The speed droop parameter of PFR
K	The integral gain of SFR
H	The inertia of the generator
K_D	The load-damping factor of the system
F_{HP}	The power fraction of the high-pressure turbine section
T_g	Time constant of the speed governor (s)
T_r	Time constant of the reheat process (s)
T_t	Time constant of the turbine (s)
Δf_W^{ACi}	Dead band of the frequency deviation (Hz)
κ_P, κ_Q	Slope parameters of the inverter AC
μ_P, μ_Q	Intercept parameters of the inverter AC
gi, gj, gk	Index of generating units (subscript)
j	Index of customers (subscript)
i	Index of aggregators (subscript)
n, p	Index of buses (subscript)
t	Index of time intervals (superscript)
C_w	Specific heat of water (J/kg $°C$)
T_s/T_r	Supply/return temperature ($°C$)
T_a/T_n	Ambient/indoor temperature ($°C$)
T_{in}/T_{out}	Inlet/outlet temperature ($°C$)
m_i	Heat water mass flow (kg/s)
H	Volumetric heat index ($W/(\mathrm{m}^3$ $°C$)
V	Peripheral volume ($W/(\mathrm{m}$ $°C$))
l	Length of the pipe (m)
h	Heat transfer coefficient ($W/(\mathrm{m}$ $°C$))
R	Relative water flow ratio
η_{es}	Efficiency of electric heating
ρ	Water density (kg/m^3)
Ngj	Number of CHP units
Ngi	Number of electricity-only units
Na	Number of aggregators
NT	Number of time intervals
B_i	Benefit function of aggregator i
$G_{e,gi}$	Output of thermal electricity-only units
$G_{e,gj}$	Power output of CHP units (MW)
$G_{h,gj}$	Heat output of CHP units (MW)
Υ_{eh}	Heat-to-electricity ratio of CHP units
L_e	Total electricity demand (MW)
L_{es}	Electricity demand (heating) (MW)
L_{e0}	Electricity demand (non-heating) (MW)

$L_{h0,j}$	Heating demand of customer j (MW)
$L_{h,i}$	Heat demand of aggregator i (MW)
G_{np}	Conductance of the transmission line
B_{np}	Susceptance of the transmission line
δ	Phase angle
V	Magnitude of the bus voltage
$T^{\max}$	Transmission line capacity (MW)
η	Efficiency of electric heating
p_e, p_h	Electricity price, heat price ($/MWh)
λ	Dual variables of the constraints
ϕ, φ	Index of thermal generating units
w	Index of wind power units
j	Index of customers
i	Index of aggregators
n, p	Index of buses
t	Index of time intervals
a, h, o	Air and mass in the house, outside air
c_w	Specific heat of water
c_a/c_h	Air/mass heat capacity
τ_s/τ_r	Supply/return temperature
$\tau_a/\tau_h/\tau_o$	Indoor air/house mass/ambient temperature
ρ_n	Static pressure of the pipeline
u	The gain/heat loss coefficient
mi,mj	Heat water mass flow
h	Heat transfer coefficient
l	Length of the pipe
η_{ES}	Efficiency of electric heating
$L_{H,\mathrm{IMP}}^{bui}$	Heat demand of the building
$L_{E,\mathrm{IMP}}^{bui}$	Electric demand of the building
$L_{E,\mathrm{ES}}^{bui}/L_{E,\mathrm{E0}}^{bui}$	Electricity demand (heating/non-heating)
$L_{H,\mathrm{ES}}^{bui}$	Heat power produced by electric heating
$L_{H,\mathrm{ALL}}^{bui}$	Total heat power injected into the building
$P_{E,i}^{dem}/P_{H,i}^{dem}$	Heat/electric demand of the aggregator i
$P_{E,n}^{dem}/P_{H,n}^{dem}$	Total heat/electric demand at bus n
$P_{E,\varphi}^{gen}$	Power output of electricity-only units
$P_{E,\phi}^{gen}\Big/P_{H,\phi}^{gen}$	Power/heat output of CHP units
$P_{H,np}^{line}$	Heat power in the heat pipeline
N_{ϕ}	Number of CHP units
N_{φ}	Number of electricity-only units
N_A	Number of customer aggregators

NT_H	Number of time intervals
R_i	Revenue function of aggregator i
C^{gen}	Generation cost function of generating unit
Υ_{HE}	Heat-to-electricity ratio of CHP units
G_{np}/B_{np}	Conductance/susceptance of the line
θ	Phase angle
V	Magnitude of the bus voltage
$T^{\max}$	Transmission line capacity
p_e, p_h	Electricity price, heat price
A/H	Area and height of the building
i, j	Index of conventional generating units
w	Index of wind power plants
l	Index of electric load
t	Index of time periods
k	Index of the state
N_C	Number of conventional generating units
N_W	Number of wind power plants
N_L	Number of electric loads
I_i	Status of the unit i (0 or 1)
c_i^u, c_i^d	Start-up/turnoff cost
R_u	Ramp rate of the generating unit
S_u	Start-up ramp rate of the generating unit
P_i^{gen}	Power out of the generating unit i
P_l^{dem}	Power demand of the electric load l
P_w^{gen}	Power out of the wind plant w
P_w^{rate}	Rated capacity of the wind plant w
$P_{w,t}^{pre}$	Day-ahead wind power forecast
$P_{w,t}^{avi}$	Real-time wind power potential
r_w	Capacity credit of the wind power plant
V_{ci}/V_{co}	Cut-in/cut-out wind speed
V_r	Rated wind speed
P_t^{disc}	Discharging power of the ESS
P_l^{inter}	Interrupted power of electric load l
P_{ESS}^{avi}	PCR of the ESS (MW/MWw)
E_{ESS}^{avi}	ECR of the ESS (MWh/MWhw)
P_{DSM}^{avi}	LCR of the ESS (%)
E_{DSM}^{avi}	LCE of the ESS (MWh/MWhw)
η_c, η_d	Charging, discharging efficiency
Δt	Interval between two time periods
p_E	Unit price of the ESS ($/MWh)
p_{DSM}	Unit price of the interrupted load ($/MWh)

E_{CGU}^{day}	Total conventional generation during a day
E_{WIND}^{day}	Total wind power generation during a day
a_i, b_i, c_i	Fuel consumption/carbon emission coefficients
μ, σ	Capacity credit of the wind power plant
σ_{total}	Expected value and standard deviation of wind speed
$\sigma_{wind}, \sigma_{load}$	Standard deviation of the wind/load forecast error

Chapter 1
Air Conditioning and Heating as Demand Response in Modern Power Systems

The utilization of renewable energy sources (RES) is burgeoning to deal with the rapidly increasing energy consumption and environment deterioration, as shown in Figs. 1.1 and 1.2 [1]. There is a consensus that the intermittency and uncertainty of wind power have been the major barriers for large scale wind power integration. The large fluctuation and severe intermittence of RES make the power generation less predictable and controllable. Furthermore, the high penetration of RES, such as wind power and photovoltaic power, has posed a great challenge to the security and reliability of the power system operation [2]. Therefore, more operating reserve is required for the system to maintain balance between power supply and demand [3]. Operating reserve is the generating capacity available in a short period of time to avoid power shortage that results from emergencies, such as random failures of the generator and load fluctuations [4]. Operating reserve is mainly provided by conventional large generators, especially thermal power generating units and hydro turbines. However, thermal power generation may be phased out in the future. Moreover, the fluctuation brought by the growing share of RES will continuously increase, while the conventional operating reserve providers may not be able to satisfy the requirements of the system with burgeoning RES in the future. Therefore, the shortage of operating reserve has become an urgent issue for both power system operation and planning [5, 6].

The development of information and communication technologies (ICT) and electricity market has made the remote control of flexible loads much easier [7]. Thus it is possible for small end-customers to provide operating reserve to support power system's operation. Studies in [8, 9] have illustrated that flexible loads have positive effects on maintaining system balance between supply and demand. Small end-customers can serve as balancing resources through the application of smart controllers and smart meters [10]. By utilizing the communication infrastructure of the smart grid, small end-customers are able to control their daily energy consumption and adapt their electricity bills to their actual economic conditions [11], as shown in Fig. 1.3.

Y. Ding et al., *Integration of Air Conditioning and Heating into Modern Power Systems*,
https://doi.org/10.1007/978-981-13-6420-4_1

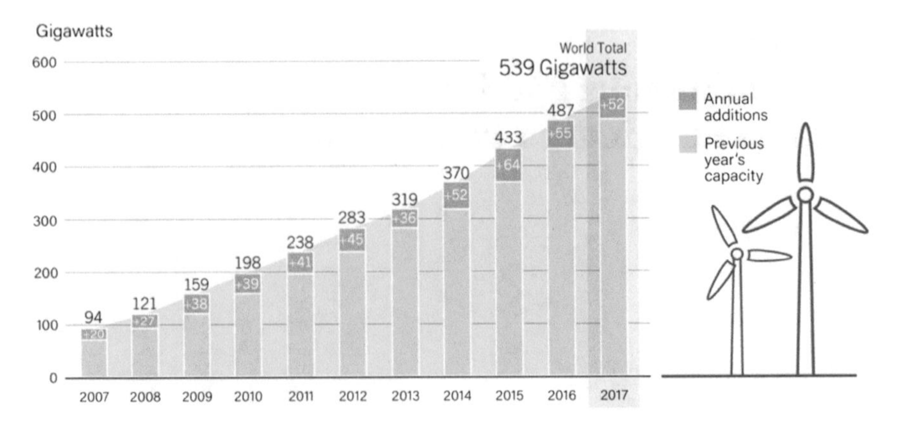

Fig. 1.1 Wind power, global capacity and annual additions, 2007–2017. (http://www.ren21.net/)

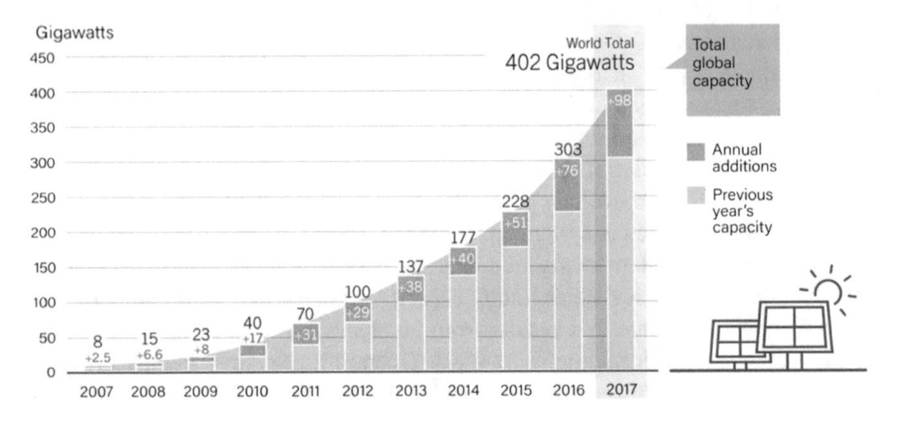

Fig. 1.2 Solar PV, global cpacity and annual additions, 2007–2017. (http://www.ren21.net/)

As one of the most popular and easily controlled flexible loads, air conditioners (AC) and heating equipments account for a large share in power consumption due to the mass application across the world [12], as shown in Fig. 1.4. According to a study carried out in Spain, electricity consumption of residential ACs accounts for about one third of the peak electricity consumption in large cities during the summer [13]. Therefore, ACs have yielded enormous potential in serving as energy storage devices, which can provide operating reserve by reducing power consumption temporarily. In this field, some researches have been conducted. For example, heating, ventilating and air-conditioning (HVAC) loads are controlled to adjust their demand profiles in response to the electricity price [7, 14, 15]. The potential for providing intra-hour load balancing services using aggregated HVAC loads has been investigated in [16]. Meanwhile, inconvenience to customers should be reduced as much as possible when ACs are controlled to provide operating reserve for power system [17–21]. Fuzzy logic-based approaches have been used in [17–19] to optimize both customer's

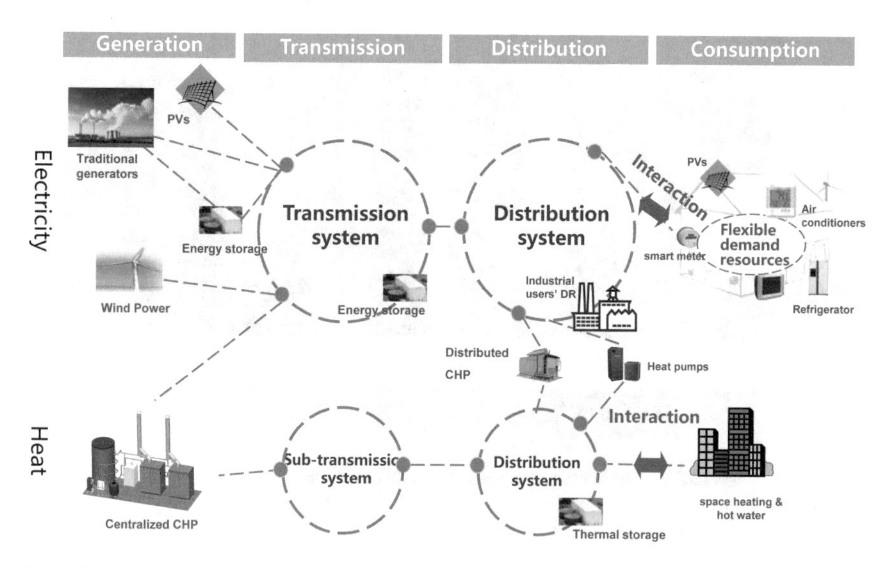

Fig. 1.3 The interaction between supply side and demand side

satisfaction and utility savings. The AC's on/off control time is considered in [20], which introduces a dynamic programming approach to minimize the load reduction in order to reduce the customers' discomfort. Reference [21] combines the advantages of linear and dynamic programming approaches to enable an acceptable level of services. Moreover, some field demonstration projects in [22–26] have shown the benefits of the demand response. For example, Con Edison, an energy company in New York City, provides customers with free smart air-conditioning kits, which help customers control their ACs remotely while earning rewards [25]. Several countries in Europe, e.g. England, Germany and Denmark, have started smart heat pump projects to help balance generation and demand [26].

Facing the huge potential of air conditioning and heating loads, this book proposes the quantitative evaluation method of the regulation service, the capacity evaluation method of aggregated thermostatically controlled loads under dynamic price signals, the sequential-dispatch of operating reserve considering lead-lag rebound effect, and the frequency regulation control method, respectively. Moreover, the integration of flexible heating demand intro the integrated energy system and a three-level framework for utilizing the demand response to improve the operation of the integrated energy system are also proposed. The economical evaluation of the flexible resources for providing the operational flexibility in the power system is also analyzed in this book.

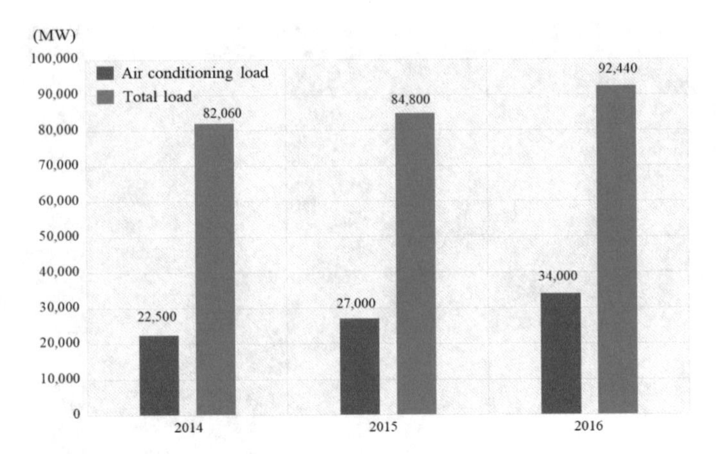

Fig. 1.4 The air conditioning loads in Jiangsu Province of China

References

1. A. Ketsetzi, M.M. Capraro, Renewable Energy Sources. A Companion to Interdisciplinary STEM Project-Based Learning (2016), pp. 145–153
2. H. Nosair, F. Bouffard, Reconstructing operating reserve: flexibility for sustainable power systems. IEEE Trans. Sustain. Energy **6**(4), 1624–1637 (2015)
3. Ž.B. Rejc, M. Čepin, Estimating the additional operating reserve in power systems with installed renewable energy sources. Int. J. Electr. Power Energy Syst. **30**(62), 654–664 (2014)
4. J. Wang, X. Wang, Y. Wu, Operating reserve model in the power market. IEEE Trans. Power Syst. **20**(1), 223–229 (2005)
5. K. Heussen, S. Koch, A. Ulbig, G. Andersson, Unified system-level modeling of intermittent renewable energy sources and energy storage for power system operation. IEEE Syst. J. **6**(1), 140–151 (2012)
6. I. Krad, E. Ibanez, W. Gao, A comprehensive comparison of current operating reserve methodologies, in *2016 IEEE/PES Transmission and Distribution Conference and Exposition (T&D)* (IEEE, 2016)
7. P. Siano, D. Sarno, Assessing the benefits of residential demand response in a real time distribution energy market. Appl. Energy **1**(161), 533–551 (2016)
8. C.L. Su, D. Kirschen, Quantifying the effect of demand response on electricity markets. IEEE Trans. Power Syst. **24**(3), 1199–1207 (2009)
9. P. Palensky, D. Dietrich, Demand side management: demand response, intelligent energy systems, and smart loads. IEEE Trans. Ind. Inform. **7**(3), 381–388 (2011)
10. Y. Ding, P Nyeng, J. Østergaard, M.D. Trong, S. Pineda, K. Kok, G.B. Huitema, O.S. Grande, Ecogrid EU-a large scale smart grids demonstration of real time market-based integration of numerous small DER and DR, in *2012 3rd IEEE PES Innovative Smart Grid Technologies Europe (ISGT Europe)*, vol. 14 (2012), pp. 1–7
11. J.M. Lujano-Rojas, C. Monteiro, R. Dufo-Lopez, J.L. Bernal-Agustín, Optimum residential load management strategy for real time pricing (RTP) demand response programs. Energy Policy **30**(45), 671–679 (2012)
12. Energy Institute at Hass, Air conditioning and global energy demand (2015). https://energyathaas.wordpress.com/2015/04/27/air-conditioning-and-global-energy-demand/
13. Science Daily, Air conditioning consumes one third of peak electric consumption in the summer (2012). https://www.sciencedaily.com/releases/2012/10/121022080408.htm

14. P. Nyeng, J. Ostergaard, Information and communications systems for control-by-price of distributed energy resources and flexible demand. IEEE Trans. Smart Grid **2**(2), 334–341 (2011)
15. G. Bianchini, M. Casini, A. Vicino, D. Zarrilli, Demand-response in building heating systems: a model predictive control approach. Appl. Energy **15**(168), 159–170 (2016)
16. N. Lu, An evaluation of the HVAC load potential for providing load balancing service. IEEE Trans. Smart Grid **3**(3), 1263–1270 (2012)
17. K. Bhattacharyya, M.L. Crow, A fuzzy logic based approach to direct load control. IEEE Trans. Power Syst. **11**(2), 708–714 (1996)
18. H.T. Yang, K.Y. Huang, Direct load control using fuzzy dynamic programming. IEE Proc. Gener. Transm. Distrib. **146**(3), 294–300 (1999)
19. H. Salehfar, P.J. Noll, B.J. LaMeres, M.H. Nehrir, V. Gerez, Fuzzy logic-based direct load control of residential electric water heaters and air conditioners recognizing customer preferences in a deregulated environment, in *Power Engineering Society Summer Meeting, IEEE 1999*, vol. 2 (1999), pp. 1055–1060
20. W.C. Chu, B.K. Chen, C.K. Fu, Scheduling of direct load control to minimize load reduction for a utility suffering from generation shortage. IEEE Trans. Power Syst. **8**(4), 1525–1530 (1993)
21. J.C. Laurent, G. Desaulniers, R.P. Malhamé, F. Soumis, A column generation method for optimal load management via control of electric water heaters. IEEE Trans. Power Syst. **10**(3), 1389–1400 (1995)
22. D.P. Chassin, D. Rondeau, Aggregate modeling of fast-acting demand response and control under real-time pricing. Appl. Energy **1**(181), 288–298 (2016)
23. W. Wang, S. Katipamula, H. Ngo, R. Underhill, D. Taasevigen, R. Lutes, Field evaluation of advanced controls for the retrofit of packaged air conditioners and heat pumps. Appl. Energy **15**(154), 344–351 (2015)
24. N. Alibabaei, A.S. Fung, K. Raahemifar, A. Moghimi, Effects of intelligent strategy planning models on residential HVAC system energy demand and cost during the heating and cooling seasons. Appl. Energy **185**, 29–43 (2016)
25. ConEdison, Control your cool with smart ACs (2016). http://www.coned.com
26. L. Sugden, Smart Grids create new opportunities for heat pumps. REHVA J., 20–22 (2012)

Chapter 2
Aggregated Air Conditioners for Providing Operating Reserve

2.1 Introduction

The exiting control strategies of air conditioners (ACs) are mostly based on the on/off control strategy, which comes into effect only by making the ACs switch between the mode of on and off. The on/off control strategy is a rough control method that sheds load directly, which will cause a sudden change in the power, bring a disturbance to the customers involved in demand response programs, and have a negative impact on the operation and performance of ACs. With the development and reform of the electric utility industry, customers' satisfaction with electric services will be increasingly more important.

Furthermore, several unified indexes, including the minimum/maximum generating capability, the start-up/shut-down time, the minimum/maximum reserve capacity, and the ramp-up/ramp-down rate limit, have been developed for evaluating the operating reserve performance provided by conventional power generating units [1, 2]. However, there are few researches which evaluate the performance of operating reserve provided by ACs for maintaining system balance. Only one index, the load-shedding capacity or load reduction, is defined to evaluate the performance of reserves for ancillary services in [3–6]. This index may not be comprehensive to evaluate the performance of operating reserve provided by ACs, since it is not clear whether the ACs can meet the requirements of providing operating reserve for the power system, how long the response time of the aggregated ACs is, and how long the load-shedding state of ACs can last. Therefore, the evaluation indexes for operating reserve provided by ACs are not sufficient. Consequently, the operating reserve provided by generation-side and demand-side cannot be dispatched collaboratively. Moreover, the lack of unified evaluation indexes makes it difficult to optimize the control of aggregated ACs, and increases the risk of unpredictable load fluctuations caused by the ACs' control.

This chapter proposes a new control strategy based on the aggregation model of ACs, as shown in Fig. 2.1. The customers, who participate in the demand response program, have signed the contract with the agreement on providing load-shedding

Y. Ding et al., *Integration of Air Conditioning and Heating into Modern Power Systems*,
https://doi.org/10.1007/978-981-13-6420-4_2

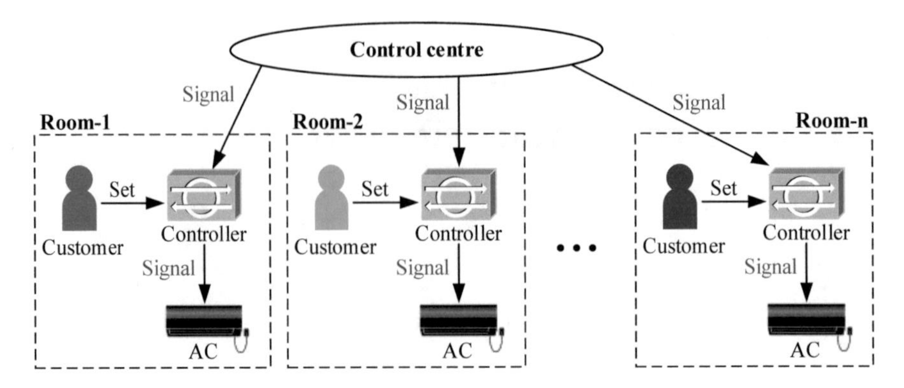

Fig. 2.1 The control structure of the aggregation model of ACs

when needed. Their houses are installed with the terminal controllers, and the customers can set the controllers' parameters, such as a target comfortable temperature and a maximum adjustable temperature deviated from the target. Accordingly, the householders can get more benefits with a larger adjustable temperature range. When power reduction is needed, the control center sends control signals (SR) to each participated terminal controller. The SR may contain: the adjustment amount of ACs' set temperature, the beginning time and the ending time of demand response. After receiving the control signal, the ACs will change the operation state, which results in reducing power consumption and providing operating reserve for that moment. Here, the adjustment amount of ACs' set temperature will be within the maximum acceptable range, which is set in advance by the customers. In this way, the customers' satisfaction can be guaranteed. Similar to the operating reserve provided by conventional generating units [7], several indexes are defined to evaluate the performance of operating reserve provided by individual AC and aggregated ACs, including reserve capacity (RC), response time (RT), duration time (DT) and ramp rate (RR). These indexes can provide an effective and efficient measurement of the maximum load-shedding capacity, the response time, the duration time, and the regulation rate.

The contributions of this chapter can be described as follows:

(a) Compared with the on/off control strategy, resetting the temperature of ACs is a softer approach to influence customers' comfort, especially when the time interval is short before adjusting to the original set temperature.
(b) Resetting the temperature of ACs will not bring a sharp drop of the power which may be adverse to the safety of the power system operation. Moreover, the ACs will be less worn and get a longer operation life by adjusting their set temperature rather than turning them on/off repeatedly.
(c) The operating reserve performance provided by individual AC and aggregated ACs is evaluated quantitatively. Several indexes are defined, including RC, RT, DT and RR, which makes up the gap of evaluating the performance of operating reserve provided by HVAC loads.

(d) Based on the proposed evaluation indexes, the operating reserve from demand-side can be dispatched by the system operator similar as conventional generating units, which will contribute to the improvement of power system's economic performance [8].
(e) The proposed evaluation indexes are calculated and analysed by simulations and case studies in this chapter. Results have shown the effectiveness of the proposed control strategy and evaluation indexes.

This chapter includes research related to the operating reserve evaluation of aggregated ACs by [9].

2.2 Operating Reserve Provided by Individual AC

2.2.1 *Thermal Model of the Room*

To study the control strategy for ACs and evaluate the performance of operating reserve provided by ACs, it is important to develop the thermal model of the room [10, 11]. The heat variation in the room is calculated by subtracting heat gains H_{gain} from heat losses H_{loss}. The temperature in the room T_A can be represented as a function of time

$$c_A \rho_A V \frac{dT_A}{dt} = H_{gain} - H_{loss} \tag{2.1}$$

where c_A, ρ_A and V denote the heat capacity of air, the density of air and the volume of the room, respectively; dT_A / dt is the temperature variation during each time interval.

The heat gains, from the AC H_{AC}, internal appliances and occupants $H_{internal}$ and the sun H_{solar}, can be expressed as

$$\begin{aligned} H_{gain} &= H_{AC} + H_{internal} + H_{solar} \\ &= P \cdot COP + \varepsilon A + P_{incident}(t) A \end{aligned} \tag{2.2}$$

where P and COP correspond to the operating power and coefficient of performance of AC, respectively; ε represents the coefficient of heat release by appliances and occupants; A is the living area of the room; $P_{incident}(t)$ is the time-varying coefficient of heat absorbed from the sun [8]. Note that H_{AC} is a positive value in the heating state and a negative value in the cooling state.

COP is an important factor for AC, which expresses the relationship between the heat supply (cooling or heating) and the power input [12]. The value of COP is related to the performance of AC's compressor, electric expansion valve, cooling load, and temperature difference between T_A and the ambient temperature T_O [13–15]. For the air-source ACs studied in this chapter, COP varies mainly with the temperature difference between T_A and T_O. Based on the data in [16], COP will be lower with a larger temperature difference, and it can be fitted to a straight line approximately, which can be expressed as

$$COP = -\theta \cdot |T_A - T_O| + \delta \tag{2.3}$$

where θ and δ are the fitted coefficients of the linear relationship between COP and $|T_A - T_O|$.

The heat losses are estimated by calculating losses through the building envelope and air leakages [8], which can be expressed as

$$H_{loss} = KA_S(T_A - T_O) + c_A\rho_A V(T_A - T_O)n \tag{2.4}$$

where K denotes the heat transfer coefficient; A_S is the surface area of the envelope; T_O denotes the ambient temperature; n is the air exchange times.

2.2.2 *Operation Characteristics of Individual Air Conditioner*

Based on the thermal model as shown above, the operation characteristics of individual AC are analysed in this subsection. It is assumed that the ACs are turned on and operated in cooling mode. The general operation characteristics of the kth individual AC are shown in Fig. 2.2, where three typical curves are presented: (a) the variation curve of the temperature in the room, (b) the state variation curve of AC and (c) the operating power variation curve of AC.

Figure 2.2a describes the variation curve of the temperature in the room, where $T_A^{(k)}$ is the temperature within the kth room. $T_{set}^{(k)}$ represents the set temperature, which is set by the customers. $T_{hy}^{(k)}$ is the hysteresis band of temperature, which describes the maximum absolute difference between the room temperature and the set temperature. Therefore, the temperature in the room varies in the range $\left[T_{set}^{(k)} - T_{hy}^{(k)}, T_{set}^{(k)} + T_{hy}^{(k)}\right]$.

The operation state of the kth AC is shown in Fig. 2.2b. ACs in summer have two operation states: cooling state $S_{cool}^{(k)}$ and standby state $S_{styb}^{(k)}$. When the temperature in the room is not higher than the lower limit value $\left(T_{set}^{(k)} - T_{hy}^{(k)}\right)$, the AC is turned to standby state. Conversely, when the temperature in the room is equal to or higher than the upper limit value $\left(T_{set}^{(k)} + T_{hy}^{(k)}\right)$, the AC is turned to cooling state. Therefore, the operation state $S^{(k)}$ can be expressed as

$$S^{(k)} = \begin{cases} S_{cool}^{(k)}, T_A^{(k)} \geq T_{set}^{(k)} + T_{hy}^{(k)} \\ S_{styb}^{(k)}, T_A^{(k)} \leq T_{set}^{(k)} - T_{hy}^{(k)} \end{cases} \tag{2.5}$$

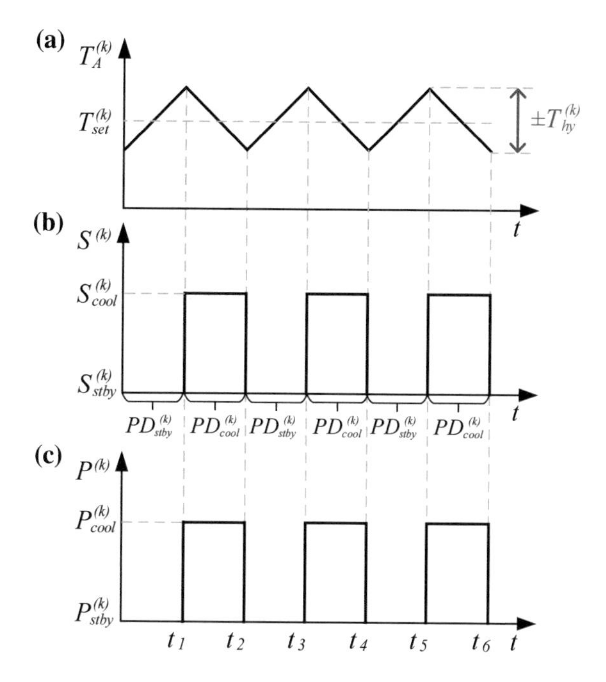

Fig. 2.2 Operation characteristics of the kth AC: **a** the variation curve of the temperature in the room, **b** the state variation curve of AC and **c** the operating power variation curve of AC

The operating power variation curve of the kth AC is similar to the state variation curve, as shown in Fig. 2.2c, where $P^{(k)}$, $P_{cool}^{(k)}$ and $P_{styb}^{(k)}$ respectively correspond to the actual operating power, the power in cooling state and the power in standby state. Therefore, it can be expressed as

$$P^{(k)} = \begin{cases} P_{cool}^{(k)}, \ S_{cool}^{(k)} \\ P_{styb}^{(k)}, \ S_{styb}^{(k)} \end{cases} \tag{2.6}$$

Conventionally, AC belongs to thermostatically controlled on/off device [1–3], which is considered as consuming constant power in cooling state and zero power in standby state.

2.2.3 *Operating Reserve Provided by Individual Air Conditioner*

Based on the operation characteristics of individual AC as shown above, this subsection analyses the operating reserve performance of individual AC. The control strategy is to reduce the power consumption by resetting AC's temperature. For instance, the set temperature of ACs in cooling state can be adjusted to a higher level to reduce power consumption, thus providing operating reserve.

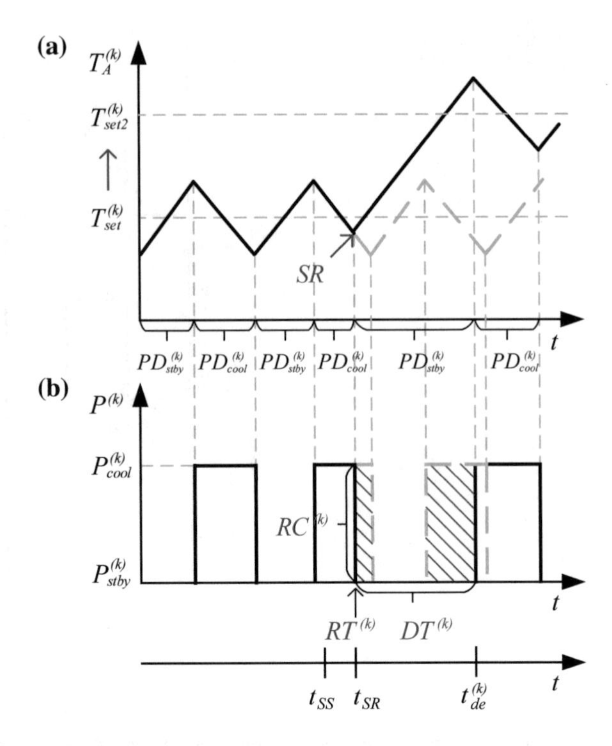

Fig. 2.3 Operating reserve provided by the kth AC when receiving the control signal in cooling state: **a** the variation curve of the temperature in the room and **b** the operating power variation curve of AC

The buildings which participate in the demand response program have been installed terminal controllers, and customers can set the controllers' parameters such as a comfortable temperature and a maximum adjustable temperature. Once power reduction is needed, the control center will send control signals, including the adjustment amount of ACs' set temperature, the beginning time and the ending time of demand response, to each terminal controller. Upon receipt of the control signal, the ACs will adjust the set temperature, which results in reducing power consumption and providing operating reserve for system operation. Here, the adjustment amount of ACs' set temperature will be in the maximum acceptable range, which is set in advance by the customers. In this way, the customers' satisfaction can be guaranteed. If before the instructed ending time, the system operator decides that the power reduction is no longer needed, a recall of the deployment can be sent and ACs can be tuned back to their original set temperature earlier than scheduled. Otherwise, original set temperature will be brought back after the instructed ending time. Compared with the on/off control strategy, resetting the temperature of ACs is a softer approach within customers' comfort zone, especially when the time interval is short before adjusting to the original set temperature.

According to the ACs' operation state when receiving the control signal, ACs' operating reserve performance can be divided into two categories: operating reserve provided by ACs in cooling state (Fig. 2.3), and in standby state (Fig. 2.4).

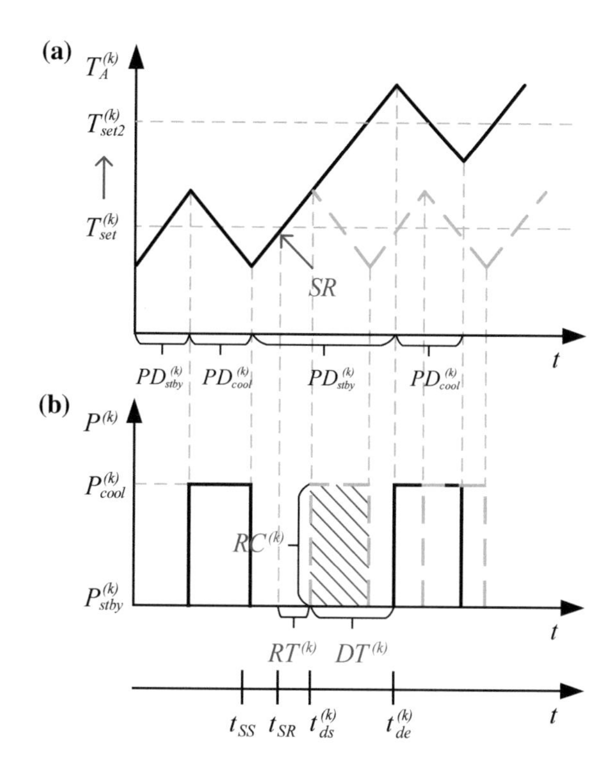

Fig. 2.4 Operating reserve provided by the kth AC when receiving the control signal in standby state: **a** the variation curve of the temperature in the room and **b** the operating power variation curve of AC

(1) Operating Reserve Provided by ACs in Cooling State

Figure 2.3 illustrates individual AC's operating reserve performance when receiving the control signal in cooling state. Figure 2.3a shows the temperature variation curve in the kth room and Fig. 2.3b shows the operating power variation curve of the kth AC. The solid line is the actual operating curve and the dashed line is the original operating curve if the set temperature is not adjusted. When the control signal is received by AC at the time t_{SR}, the set temperature of the AC is reset from $T_{set}^{(k)}$ to $T_{set2}^{(k)}$. Then the AC turns to standby state and keeps in that state until the temperature in the room rises to the upper limit value of $\left(T_{set2}^{(k)} + T_{hy}^{(k)}\right)$. Therefore, ACs in cooling state can reduce power consumption and provide operating reserve in a fast response manner.

(2) Operating Reserve Provided by ACs in Standby State

Figure 2.4 illustrates individual AC's operating reserve performance when receiving the control signal in standby state. Figure 2.4a is the temperature variation curve in the kth room and Fig. 2.4b is the operating power variation curve of the kth AC. Although the ACs in standby state have no power loads, these ACs will also reset the temperature from $T_{set}^{(k)}$ to $T_{set2}^{(k)}$ after receiving the control signal. Then the standby time will be extended to the time $t_{de}^{(k)}$. If the set temperature is not adjusted, the AC will turn to cooling state at the time $t_{ds}^{(k)}$. It is equivalent that ACs in standby state start to provide reserve capacity at the time $t_{ds}^{(k)}$ when the AC is supposed to work.

In order to evaluate the performance of operating reserve provided by individual AC, several indexes are defined, including reserve capacity (RC), response time (RT) and duration time (DT). RC is the maximum reserve capacity provided by the AC, which is equal to its cooling power. Therefore, the reserve capacity provided by ACs in both states can be expressed as

$$RC^{(k)} = P_{cool}^{(k)} \tag{2.7}$$

where $RC^{(k)}$ and $P_{cool}^{(k)}$ represents the reserve capacity and the cooling power of the kth AC, respectively.

Response time (RT) is the time delay before the AC starts to provide operating reserve after the control signal is sent. For ACs in cooling state, RT is the control signal communication time. While for ACs in standby state, RT has to add the original standby time, which can be expressed as

$$RT^{(k)} = \begin{cases} t_{SR} - t_{SS}, t_{SR} \in PD_{cool}^{(k)} \\ t_{ds}^{(k)} - t_{SS}, t_{SR} \in PD_{styb}^{(k)} \end{cases} \tag{2.8}$$

where $PD_{cool}^{(k)}$ and $PD_{styb}^{(k)}$ are the time periods of cooling state and standby state, respectively.

Duration time (DT) is the length of time that the AC keeps in standby state until it restarts to refrigerate. DT of ACs in both states can be defined as

$$DT^{(k)} = \begin{cases} t_{de}^{(k)} - t_{SR}, t_{SR} \in PD_{cool}^{(k)} \\ t_{de}^{(k)} - t_{ds}^{(k)}, t_{SR} \in PD_{styb}^{(k)} \end{cases} \tag{2.9}$$

2.3 Operating Reserve Provided by Aggregated ACs

Based on the operating reserve performance of individual AC as shown above, the operating reserve provided by the aggregated ACs is analysed in this section. The aggregated model contains multi-ACs, which have different rated powers and different coefficients of performance. The set temperature and operation state of these ACs are also different. Thus the reserve capacity provided by aggregated ACs is time-varying as the result of different response time and capacity of each AC.

Typical operating reserve provided by the aggregated ACs is shown in Fig. 2.5, where P_{max} and P_{min} correspond to the maximum power and minimum power of the aggregated ACs, respectively. The control signal is sent by the control center at the time t_{SS}. Because of the different response time, the ACs start to respond and provide reserve one by one rather than all at once. The aggregated ACs have the

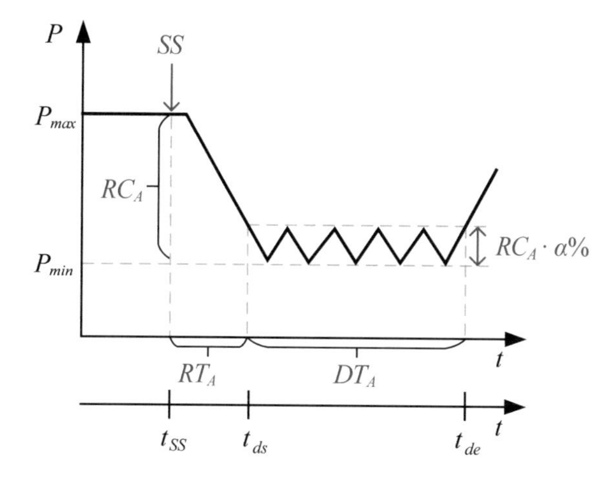

Fig. 2.5 Operating reserve provided by aggregated ACs

minimum power consumption at the time t_{ds} and start to provide operating reserve. The aggregated ACs finish providing operating reserve at the time t_{de} when the power consumption of the aggregated ACs are rising. Because the minimum power is not a strict horizontal line, $\alpha\%$ is a valid interval from the maximum load-shedding capacity to a certain range, and $(RC_A \cdot \alpha\%)$ is regarded as the valid range in which the aggregated ACs provide operating reserve. And $(t_{de} - t_{ds})$ is regarded as the duration time of providing operating reserve.

2.3.1 Performance of Operating Reserve Provided by Aggregated ACs

Several indexes are proposed in this chapter to evaluate the performance of operating reserve provided by aggregated ACs, including reserve capacity (RCA), response time (RTA), duration time (DTA) and ramp rate (RRA).

(1) **Reserve Capacity**

Different from the reserve capacity provided by individual AC, RCA is defined as a valid range around the maximum reserve capacity, since the total power of aggregated ACs is fluctuating, even when it has reached the maximum shedding capacity.

It is assumed that the operating power of ACs at the time t_{SS} is P_{max}, and the minimum operating power is P_{min}. Therefore, the reserve capacity of the aggregated ACs can be represented as

$$RC_A = P_{max} - P_{min} \tag{2.10}$$

(2) **Response Time and Duration Time**

RTA is the length of time from the moment when the control signal is sent to the moment when valid reserve capacity is achieved. DTA is the length of time when the reserve capacity is within the valid range of RCA.

As shown in Fig. 2.5, t_{ds} and t_{de} can be evaluated according to the intersections of the operating power curve $P(t)$ and the upper boundary of the reserve capacity's valid range ($P_{min} + RC_A \cdot \alpha\%$). The intersections can be achieved by

$$P(t) = P_{min} + RC_A \cdot \alpha\% \tag{2.11}$$

Based on the two solutions t_{ds} and t_{de}, the response time and duration time can be calculated as

$$RT_A = t_{ds} - t_{SS} \tag{2.12}$$

$$DT_A = t_{de} - t_{ds} \tag{2.13}$$

(3) **Ramp Rate**

RRA is the ratio of RCA and RTA, which reflects the rate of providing reserve capacity by aggregated ACs. According to the reserve capacity and response time, the ramp rate can be calculated as

$$RR_A = RC_A \cdot (1 - \alpha\%) / RT_A \tag{2.14}$$

2.3.2 *Simulation Framework for Evaluating Operating Reserve Performance*

Based on the thermal model of the room and the control strategy of AC, the operating reserve performance of the aggregated ACs can also be simulated.

Figure 2.6 shows the detailed flow chart of the simulation, which can be described by the following steps:

(a) Initialize the parameters, including AC parameters (such as rated powers, set temperatures and COP), room parameters (such as living areas and specific heat transfer coefficients of the building envelop) and ambient temperature.
(b) Determine whether ACs have been turned on. The turned off ACs are not available, and the program jumps to step g.
(c) Search for the control signal. The ACs will reset the set temperature from $T_{set}^{(k)}$ to $T_{set2}^{(k)}$ after receiving the signal. Otherwise, the set temperature will remain unchanged.
(d) Calculate the power of the kth AC and the corresponding heat flow produced by the AC.

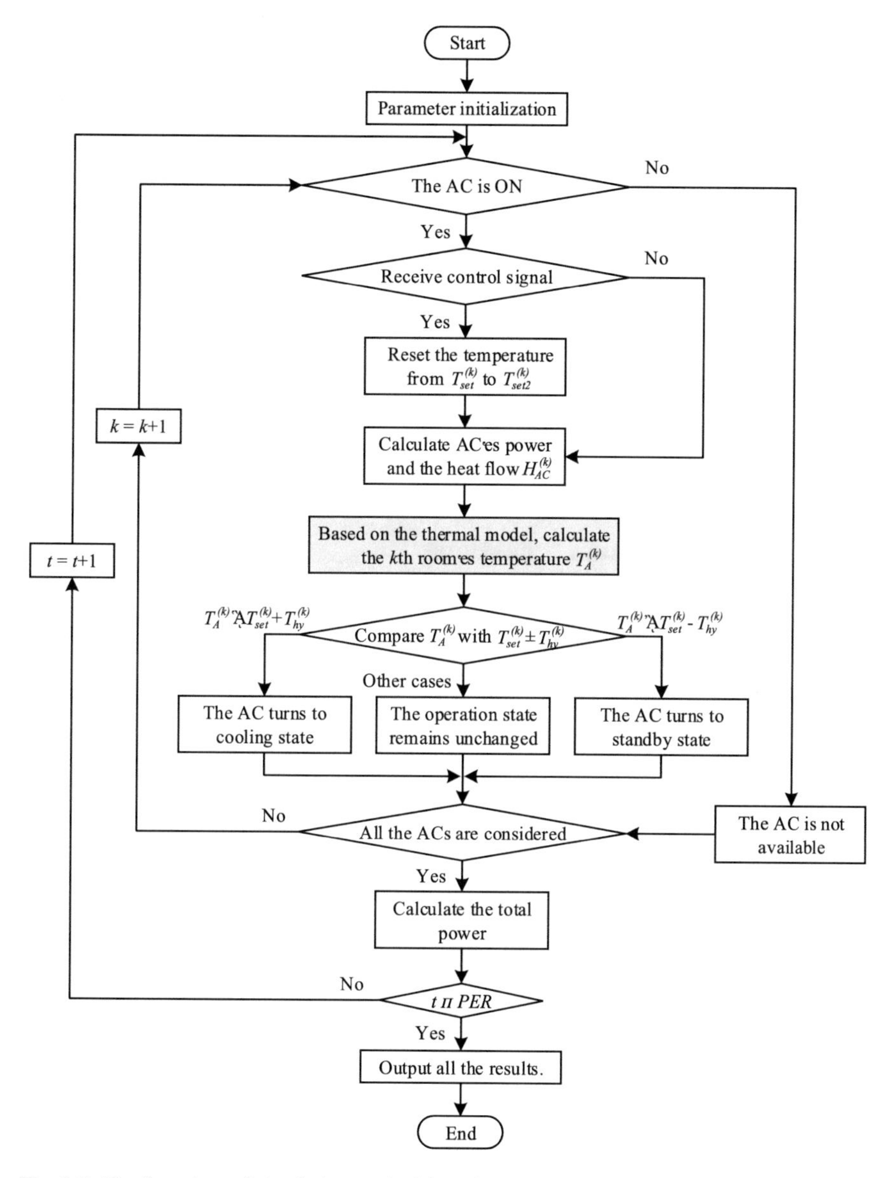

Fig. 2.6 The flow chart of simulation method for calculating operating reserve

(e) Based on the thermal model of the room, calculate the kth room's temperature $T_A^{(k)}$ at the time t.
(f) Compare $T_A^{(k)}$ with the set temperature. The AC will either turn to cooling state if $T_A^{(k)} \geq T_{set}^{(k)} + T_{hy}^{(k)}$, or turn to standby state if… And the operation state remains unchanged in other cases.
(g) Determine whether all the ACs have been considered. If some ACs have not been calculated, the program will loop from step b to step g. Otherwise, the program jumps to step h.
(h) Calculate the total power at the time t.
(i) Determine whether the simulation period PER has been finished. If the time t is smaller than PER, the program will loop from step b to step i. Otherwise, the program jumps to step j.
(j) Output all the results.

2.4 Case Studies and Discussions

This section evaluates the operating reserve performance of aggregated ACs by representative cases. It is organized as parameter initialization, operating reserve performance with different temperature adjustments, operating reserve performance with different numbers of ACs, analysis of aggregated ACs returning to original set temperature and analysis of demand response in actual case studies.

2.4.1 Parameter Initialization

Some fixed parameters are shown in Table 2.1 [16–18]. The variable parameters are set as follows.

The number of ACs is N, which is equal to the number of corresponding rooms.

The living area of the N rooms are generated in the normal distribution by using the mean value of 100 m^2 and the standard deviation of 40 m^2.

The initial set temperatures of ACs distribute randomly between 23 and 26 °C.

The rated power of each AC is related to the living area of the corresponding room. In general, the rated power will be higher in a bigger room. Here it is assumed that each rated power equals to 60-fold living area. For example, the rated power is 1800 W if the room area is 30 m^2.

The ambient temperature is the actual monitored data of a city, on August 1, 2015 [19], as shown in Fig. 2.7. And the control signal sending time is 12:00 AM.

Table 2.1 Fixed parameter initialization [16–18]

Symbols	Definitions or descriptions	Values	Units
h	Height of the room	2.5	m
θ	Fitted coefficient of COP	0.0384 [16]	N/A
δ	Fitted coefficient of COP	3.9051 [16]	N/A
K	Heat transfer coefficient	7.69 [16]	W/°C m^2
C_A	Heat capacity of air	1.005 [17]	kJ/(kg °C)
ρ_A	Density of air	1.205 [17]	kg/m^3
ε	Coefficient of heat release by appliances and occupants	4.3 [18]	W/m^2
n	Air exchange times	0.5	1/h
$T_{hy}^{(k)}$	Hysteresis band of temperature control	± 1	°C
$\alpha\%$	Valid interval of providing operating reserve	10%	N/A

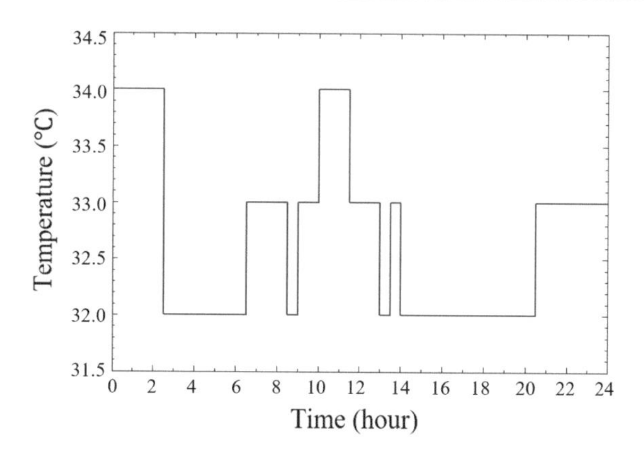

Fig. 2.7 The ambient temperature [19]

2.4.2 *Operating Reserve Performance with Different Temperature Adjustments*

This section will analyze the operating reserve performance with different temperature adjustments. The number of ACs is set to 100 and three conditions of different adjustment amounts of the set temperature ΔT_{set} are considered: $\Delta T_{set} = 1$ °C, $\Delta T_{set} = 2$ °C and $\Delta T_{set} = 3$ °C.

To directly demonstrate the operating process of the aggregated ACs, the temperature and power variation processes of 100 aggregated ACs with different temperature adjustments are shown in Fig. 2.8 and Fig. 2.9, respectively. Figure 2.8 indicates that the temperature of the 100 aggregated ACs are time-varying after adjusting the set

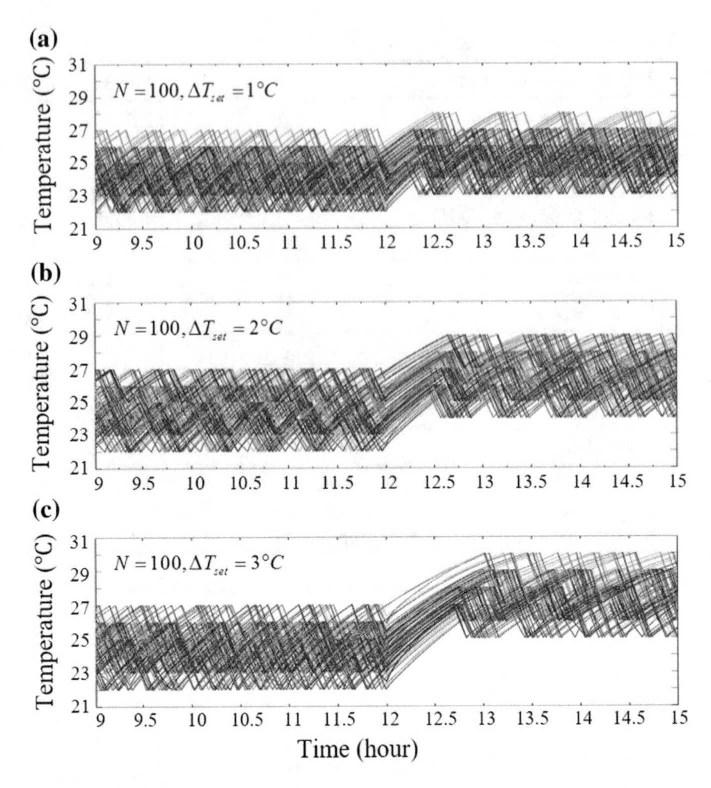

Fig. 2.8 Temperature variation curves of 100 ACs with different temperature adjustments, **a** 1 °C, **b** 2 °C and **c** 3 °C

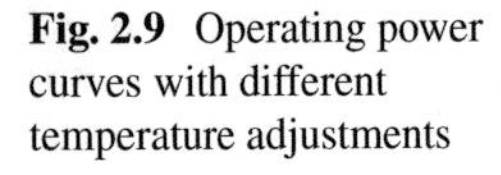

Fig. 2.9 Operating power curves with different temperature adjustments

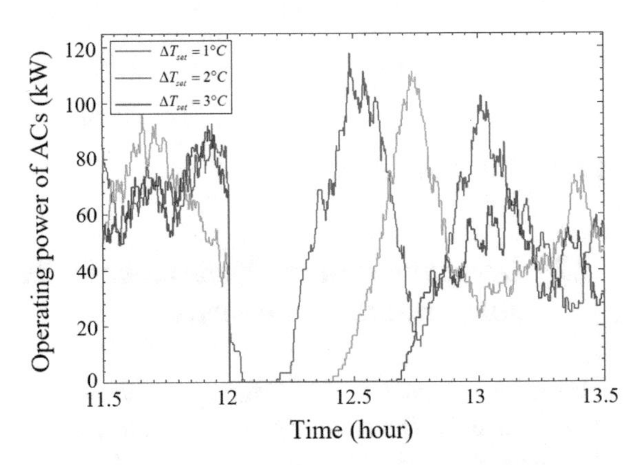

temperature. The power curve in Fig. 2.9 is similar to the theoretical power curve shown in Fig. 2.5, which validates the effectiveness of the control strategy of the aggregated ACs.

Table 2.2 Four indexes in three kinds of temperature adjustments

Indexes	$\Delta T_{set} =$ 1 °C	$\Delta T_{set} =$ 2 °C	$\Delta T_{set} =$ 3 °C	Units
RCA	92.63	96.06	92.01	kW
RTA	2.83	0.21	0.16	Min
RRA	32.73	457.43	575.06	kW/Min
DTA	12.67	28.25	40.42	Min

The proposed indexes, reserve capacity (RCA), response time (RTA), duration time (DTA) and ramp rate (RRA), are calculated and shown in Table 2.2.

Several observations can be made from above simulation results:

AC aggregation can provide operating reserve, and the maximum reserve capacity can be reached in a short time.

Reserve capacities in the three cases are almost the same. When the number of ACs remains the same, the total installed power capacity will be roughly similar. Therefore, although the temperature adjustments are different, reserve capacities which AC aggregation can achieve are almost the same.

Response time in the three cases are all within 5 min, and the time will decrease along with the increase of the temperature adjustments. Especially, when the adjustment amount of temperature is more than 2 °C, the response can be accomplished instantaneously. Response time will be a little longer in practice, due to the fact that the communication time of control signal is neglected in the simulation. However, the communication technology is able to achieve second-level communication nowadays. Therefore, the communication impact on response time is not significant.

Ramp rate is the ratio of reserve capacity to the response time. The value will be very large when the response time is relatively small. For example, ramp rate is 575.06 kW/Min when the temperature adjustment is 3 °C.

Duration time in the three cases are extended along with the increase of the temperature adjustments. The more set temperatures are adjusted, the longer standby time will last, hence leading to a longer duration time. The relationship between duration time and the adjustment amount of the set temperatures can be fitted as a linear function, as shown in Fig. 2.10.

2.4.3 *Operating Reserve Performance with Different Numbers of ACs*

This subsection will analyze the operating reserve performance under different numbers of ACs. The number is set to 100, 500, 1000 and 2000. The adjustment amount of the set temperature ΔT_{set} is 1 °C. The temperature and power variation processes of aggregated ACs are shown in Fig. 2.11 and Fig. 2.12, respectively.

The proposed four indexes can be calculated, as shown in Table 2.3.

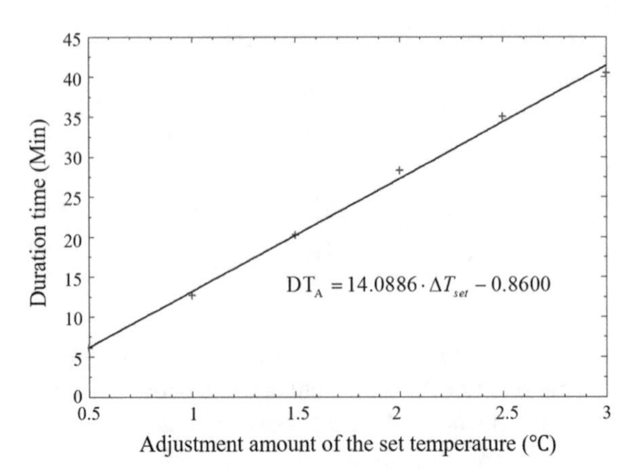

Fig. 2.10 Fitted curve of duration time and temperature adjustments

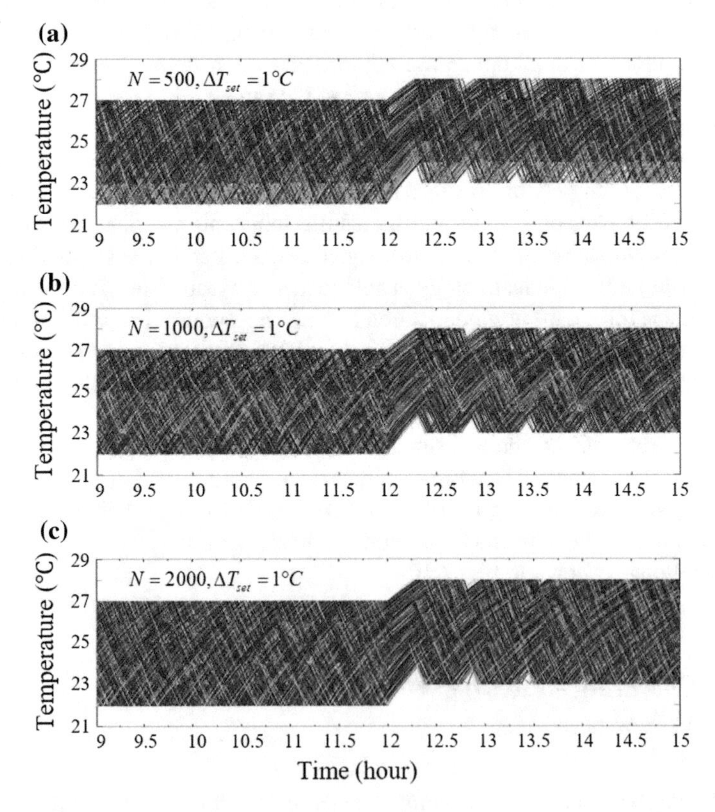

Fig. 2.11 Temperature variation curves with different numbers of ACs, **a** 500, **b** 1000 and **c** 2000

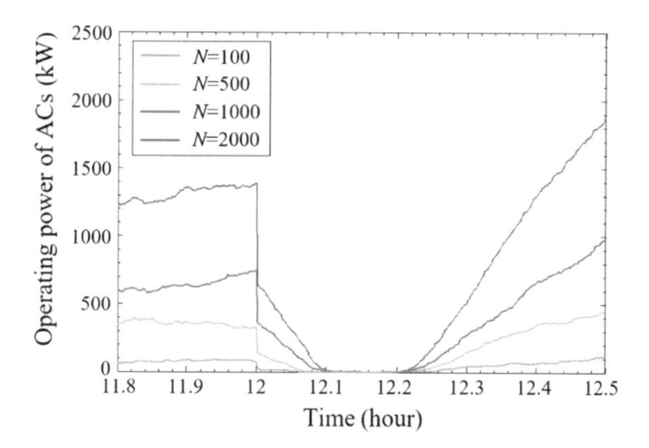

Fig. 2.12 Operating power curves with different numbers of ACs

Table 2.3 Four indexes with different number of ACs

Indexes	$N = 100$	$N = 500$	$N = 1000$	$N = 2000$	Units
RCA	92.63	388.89	789.10	1578.00	kW
RTA	2.83	3.00	4.08	4.33	Min
RRA	32.73	129.63	193.41	364.43	kW/Min
DTA	12.67	12.42	11.17	10.58	Min

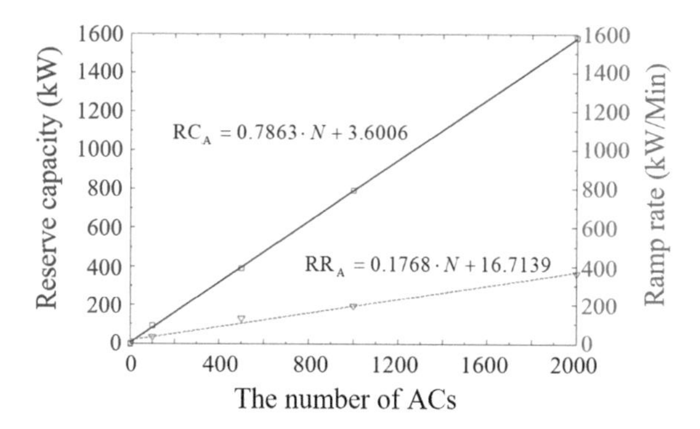

Fig. 2.13 Fitted curve of reserve capacity and ramp rate with different numbers of ACs

Compared with the results in Sect. 2.4.2, the simulation illustrates that the response time (RTA) and duration time (DTA) are most dependent on temperature adjustment, not on the number of ACs. However, reserve capacity (RCA) and ramp rate (RRA) will increase along with the number of ACs. The variation tendency of the two indexes with the number N can be described in Fig. 2.13.

The curve-fitting solid line and dashed line respectively correspond to the reserve capacity and ramp rate, which are directly proportional to the number of ACs.

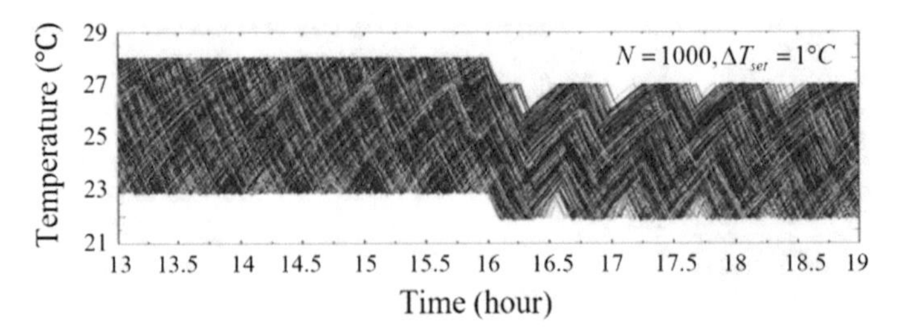

Fig. 2.14 Temperature variation curves of aggregated ACs returning to original set temperature

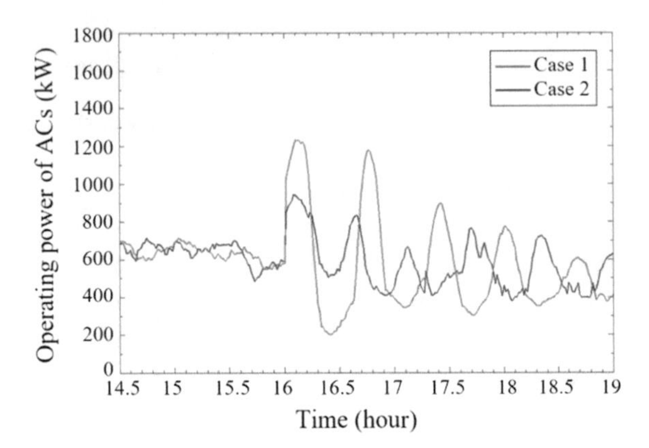

Fig. 2.15 Operating power curves of aggregated ACs returning to original set temperature

2.4.4 *Analysis of Aggregated ACs Returning to Original Set Temperature*

This subsection will analyse the operating performance of aggregated ACs returning to original set temperature. The studied number of ACs is 1000 and the adjustment amount of the set temperature ΔT_{set} is set to 1 °C. Demand response ends at 16:00, and each AC returns to their original set temperature from $T_{set2}^{(k)}$ to $T_{set}^{(k)}$. The temperature and power variation processes of aggregated ACs are shown in Fig. 2.14 and Fig. 2.15, respectively.

There are two cases in Fig. 2.15. In Case 1, all the aggregated ACs return to original set temperature at 16:00. A power rebound occurs, because all the ACs will turn to cooling state at almost the same time. Due to the thermal insulation properties of each building are with little difference in the same region, the duration time of each AC in cooling state or standby state is similar. Therefore, the power oscillation lasts for a lengthy period of time in this simulation [20].

In order to reduce the sharp rebound of power, a batch returning method is simulated in Case 2. ACs are divided into five groups. Every half an hour, a group of ACs returns to original set temperature. The second curve in Fig. 2.15 shows that power oscillation is confined to a certain range.

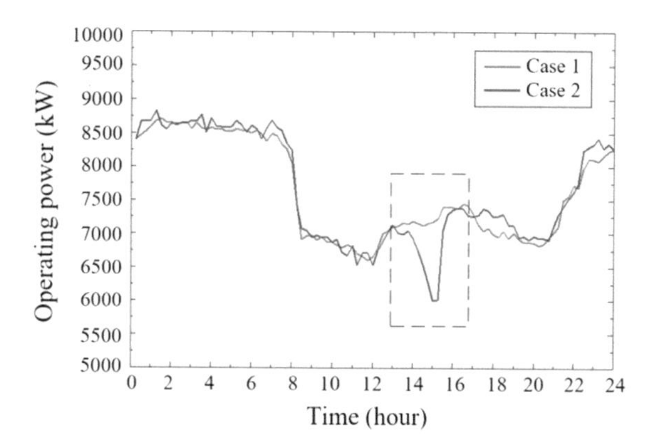

Fig. 2.16 Operating power curves in actual case studies

2.4.5 Analysis of Demand Response in Actual Case Studies

ACs can lead to the load peak and cause stress on the power system especially in hot days. Moreover, the increasing penetration of RES, such as wind power and photovoltaic power, brings more fluctuation and intermittence to the power generation. In order to testify the effectiveness of the proposed control strategy and evaluate the reserve capacity performance of residential ACs, a case study has been carried out by applying the proposed control strategy to real-recorded household demand through a demonstration project in a province of China, where the number of ACs rises rapidly, and their electricity consumption has accounted for more than 30 percent in summer. Each household participating in the project is equipped with a smart meter and a terminal energy controller. In addition, the householders sign the contract regarding how many times that they agree to be controlled in a year and how they will be compensated after demand response event.

There are 522 households participating in the demand response program, whose electricity consumption are collected every 15 min for two weeks. Two cases are considered:

Case 1: There is no demand response signal sent to the customers' terminal controller in the first week. The original power consumption curve is calculated by the average of the seven days' electricity consumption data.

Case 2: There is a beginning signal of demand response sent to the terminal controllers at 14:00 and an ending signal at 15:00 every day during the second week. Similarly, the second power consumption curve is calculated by the average of the seven days' electricity consumption data.

The weather condition is similar during the selected two weeks. And it is assumed that these customers' electric-equipment do not change. The results are shown in Fig. 2.16.

It can be seen that the demand response program performed in the second week causes peak load shifting: the load begins to decrease at 14:00 and reaches the lowest point at 14:56. The maximum power reduction is 1218.12 kW, which is the reserve capacity (RCA). The power begins to rebound at 15:17 and returns to the normal level at 15:45. Base on the evaluation method of operating reserve, the other three indexes can be calculated. The response time (RTA), ramp rate (RRA) and duration time (DTA) are around 33 min, 40.61 kW/Min and 48 min, respectively. It demonstrates that demand response can provide operating reserve by reducing power consumption.

2.5 Conclusions

This chapter proposes a method to quantitatively analyze the operating reserve performance of the aggregated ACs. First, a softer control method on ACs is proposed and several indexes on the performance of operating reserve provided by AC aggregation are defined. These indexes, including reserve capacity, response time, duration time and ramp rate, are similar to the evaluation indexes of conventional power generations. Moreover, a simulation method on evaluating these indexes is proposed. The case study results show that AC aggregation can reach the maximum load-shedding within 5 min, and meet the requirements for providing operating reserve. The simulation results and the demonstration program validate the effectiveness of the proposed evaluation indexes for operating reserve provided by aggregated ACs.

Operating reserve capacity provided by ACs mainly relates to two factors, acceptable temperature adjustments and the number of ACs. Response time and duration time are mainly affected by temperature adjustments. Response time will be shorter or even instantaneous, when the temperature adjustment is large enough. Duration time will extend significantly along with the increase of the temperature adjustment. However, the other two indexes, reserve capacity and ramp rate will increase in direct proportion with the ACs' number. Furthermore, customer's comfort will be affected by a larger temperature adjustment. Therefore, in order to improve the potential of operating reserve, promoting more customers to participate in the demand response program is important.

This chapter has made up the gap of evaluating the performance of operating reserve provided by HVAC loads. Based on the indexes proposed in this chapter, the operating reserve from demand-side can be dispatched by the system operator as conventional generating units, which is a novel and meaningful practice for the power system. Moreover, this research can guide the demand response program and improve power system's economy.

References

1. C. Wang, S.M. Shahidehpour, Effects of ramp-rate limits on unit commitment and economic dispatch. IEEE Trans. Power Syst. **8**(3), 1341–1350 (1993)
2. Y. Ding, L. Cheng, Y. Zhang, Y. Xue, Operational reliability evaluation of restructured power systems with wind power penetration utilizing reliability network equivalent and time-

sequential simulation approaches. J. Mod. Power Syst. Clean Energy **2**(4), 329–340 (2014)
3. Z. Csetvei, J. Østergaard, P. Nyeng. Controlling price-responsive heat pumps for overload elimination in distribution systems, in *2011 2nd IEEE PES International Conference and Exhibition on Innovative Smart Grid Technologies (ISGT Europe), 2011 Dec 5* (2011), pp. 1–8
4. K. Bhattacharya, Competitive framework for procurement of interruptible load services. IEEE Trans. Power Syst. **18**(2), 889–897 (2003)
5. Y.C. Huang, Integrating direct load control with interruptible load management to provide instantaneous reserves for ancillary services. IEEE Trans. Power Syst. **19**(3), 1626–1634 (2004)
6. N. Lu, An evaluation of the HVAC load potential for providing load balancing service. IEEE Trans. Smart Grid **3**(3), 1263–1270 (2012)
7. J. Wang, X. Wang, Y. Wu, Operating reserve model in the power market. IEEE Trans. Power Syst. **20**(1), 223–229 (2005)
8. D. Ryder-Cook, Thermal modelling of buildings. Cavendish Laboratory, Department of Physics, University of Cambridge, Technical Report 2009 May 11
9. H. Hui, Y. Ding, D. Liu, Y. Lin, Y. Song, Operating reserve evaluation of aggregated airconditioners. Appl. Energy **196**, 218–228 (2017)
10. J.F. Straube, E.F. Burnett, *Building Science for Building Enclosures* (Building Science Press, 2005)
11. M.L. Zheng, R.Y. Fang, Z.T. Yu, Life cycle assessment of residential heating systems: a comparison of distributed and centralized systems, in *Applied Energy Symposium and Forum 2016: Low Carbon Cities & Urban Energy Systems* (2016)
12. J. Ji, T.T. Chow, G. Pei, J. Dong, W. He, Domestic air-conditioner and integrated water heater for subtropical climate. Appl. Therm. Eng. **23**(5), 581–592 (2003)
13. Y.C. Park, Y.C. Kim, M.K. Min, Performance analysis on a multi-type inverter air conditioner. Energy Convers. Manag. **42**(13), 1607–1621 (2001)
14. S. Shao, W. Shi, X. Li, H. Chen, Performance representation of variable-speed compressor for inverter air conditioners based on experimental data. Int. J. Refrig. **27**(8), 805–815 (2004)
15. D. Qv, B.B. Dong, L. Cao, L. Ni, J.J. Wang, R.X. Shang, Y. Yao, An experimental and theoretical study on an injection-assisted air-conditioner using R32 in the refrigeration cycle. Appl. Energy. Available online 9 Nov 2016
16. F. Fang, Simulation and impact studies of Ecogrid EU reserve. Diss. Master's thesis, Technical University Denmark, Lyngby, Denmark (2013)
17. J.R. Zhang, T.Y. Zhao, *Handbook of Thermo Physical Properties of Common Substances in Engineering* (National Defense Industry Press, Beijing, 1987)
18. Design standard for energy efficiency of residential buildings in hot summer and cold winter zone. JGJ134-2010. Ministry of Housing and Urban-Rural Development of the People's Republic of China (2010)
19. Weather Underground, Hourly Weather History & Observations in Hangzhou (2015). https://www.wunderground.com/cgi-bin/findweather/getForecast?query=hangzhou
20. N.A. Sinitsyn, S. Kundu, S. Backhaus, Safe protocols for generating power pulses with heterogeneous populations of thermostatically controlled loads. Energy Convers. Manag. **67**, 297–308 (2013)

Chapter 3
Heterogeneous Air Conditioner Aggregation for Providing Operating Reserve Considering Price Signals

3.1 Introduction

As illustrated in Chaps. 1 and 2, the increasing penetration of renewables brings more fluctuations to electric power systems [1]. Therefore, the requirement of operating reserve capacity (ORC) for maintaining power balance between supply and demand is larger [2–4]. Information and communication technology have developed, which makes it possible for demand side resources (DSRs) to provide operating reserve by reducing or shifting loads [5, 6]. Thermostatically controlled loads (TCLs), such as heating, ventilation and air conditioning, account for a large share of power consumptions. For example, the power proportion of residential air conditionings reaches up to 40% during summer peak load periods in China [7, 8]. Moreover, consumer's comfort will not be affected when the operating states of TCLs are adjusted temporarily [9, 10]. Therefore, TCLs have a great potential to be controlled and provide ORC [11]. A TCL model is developed in [12] to participate in operating reserve services. Reference [13] proposes centralized control methods on TCLs to provide operating reserve. The comparison of distributed system and centralized system is studies in [14]. Besides, a load following method is developed in [6] to enhance the safety and stability of the power system. An operational planning framework for aggregated TCLs is developed to improve the efficiency in day-ahead scheduling and real-time operation [15, 16].

Price-based demand response (DR) is one of the main approaches for DSRs providing operating reserve services [10, 17]. Consumers can adjust the power consumption to respond the variable electricity prices [18] and reduce their electricity expenditure [19, 20]. Moreover, the social welfare is improved based on the price-based DR [21] and the optimization mechanism [22, 23]. However, two practical problems of ORC evaluation are relatively less studied: On the one hand, the lack of consumer behavior model makes it difficult to evaluate the ORC provided by aggregated TCLs. Consumers have diverse preferences on power consumption when the electricity prices are fluctuated [24–26]. Therefore, the consumer decision making process is relatively vague. On the other hand, it is impractical to obtain all the hetero-

Y. Ding et al., *Integration of Air Conditioning and Heating into Modern Power Systems*,
https://doi.org/10.1007/978-981-13-6420-4_3

geneous parameters, especially for large-scale TCL aggregations. Therefore, ORC has to be evaluated based on insufficient data of aggregated heterogeneous TCLs.

In this chapter, an ORC evaluation method for large-scale aggregated heterogeneous TCLs is proposed based on insufficient measurement data. Firstly, an individual TCL model is developed to evaluate ORC provided by a consumer. Then, satisfaction index is quantified by the fuzzy set method. Consumer's decision-making process and behaviors are simulated with the aim of maximizing satisfactions. Finally, the probability density estimation (PDE) method is proposed to evaluate ORC with insufficient data. The main contributions of this chapter can be summarized as follows:

(a) The individual TCL model integrating electric-thermal characteristics and consumer behaviors, is developed for providing ORC considering consumer satisfaction with price signals, which has been rarely studied in the existing literatures.
(b) Consumer preferences on the electricity price and the room temperature are modeled with the fuzzy set method. In this manner, consumer's cognition and trade-off in decision making process can be quantified for consumer behavior simulation.
(c) The PDE method is proposed to evaluate the ORC of large-scale heterogeneous TCLs without sufficient measurement data. Compared with the traditional moment estimation method, the evaluation precision of ORC is improved significantly.

This chapter includes research related to the operating reserve capacity evaluation of aggregated heterogeneous TCLs with price signals by [27].

3.2 Individual TCL Model

3.2.1 Framework and Electric-Thermal Model

To evaluate ORC provided by aggregated TCLs, the framework of an individual TCL model is proposed, as shown in Fig. 3.1. The individual TCL model comprises the consumer model and the electric-thermal model, the latter of which is divided into electric model of the TCL and the thermal model of a room. Moreover, the input and output of the individual TCL model are price signals and ORC, respectively [3, 28].

The consumer model is developed to simulate the consumer behaviors, which considers the electricity cost and room temperature. Based on the preference of the two factors, index of consumer's satisfaction is defined. Then, optimal control strategy is designed to maximize consumers' satisfaction level. The modeling details will be discussed in the next two subsections.

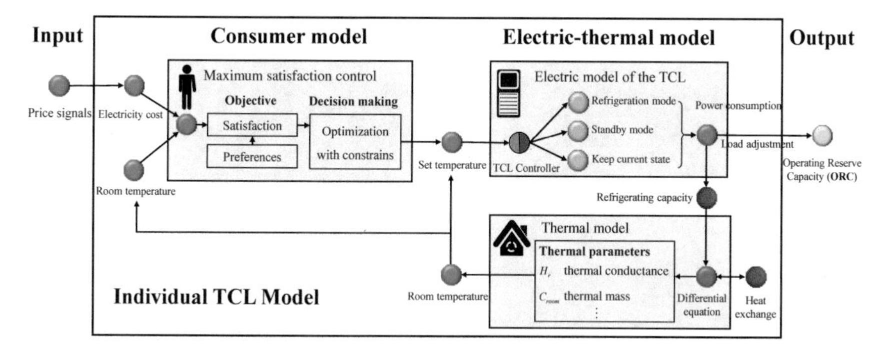

Fig. 3.1 The framework of the individual TCL model

It is assumed that TCLs work in refrigeration mode during summer period. The operating state of the TCL is decided by the set temperature $T_{set}(t)$ and the current room temperature $T_{room}(t)$, which can be expressed as

$$S(t)=\begin{cases} 1,\ T_{room}(t) \geq T_{set}(t) + \Delta T \\ 0,\ T_{room}(t) \leq T_{set}(t) - \Delta T \\ S(t-\tau),\ else \end{cases} \tag{3.1}$$

where $S(t)$ is the operating state of the TCL; ΔT represents the hysteresis band for room temperature control; τ is the time interval of each control. The TCL will turn to refrigeration mode ($S(t) = 1$) if the room temperature is higher than the set temperature, while the TCL will turn to standby mode ($S(t) = 0$) if the room temperature is lower than the set temperature.

The power consumption and the electricity cost of the TCL can be expressed as

$$P(t) = P_r \cdot S(t) \tag{3.2}$$

$$C(t) = P(t) \cdot p(t) \tag{3.3}$$

where $P(t)$ and P_r is the power consumption and the rated power of the TCL, respectively; $C(t)$ is the electricity cost; $p(t)$ is the electricity price.

Moreover, the refrigerating capacity provided by the TCL can be expressed as

$$Q(t) = EER \cdot P(t) \cdot \tau \tag{3.4}$$

where EER is the energy efficiency ratio between the power consumption and the refrigerating capacity.

The thermal model of the room can be described as [29]

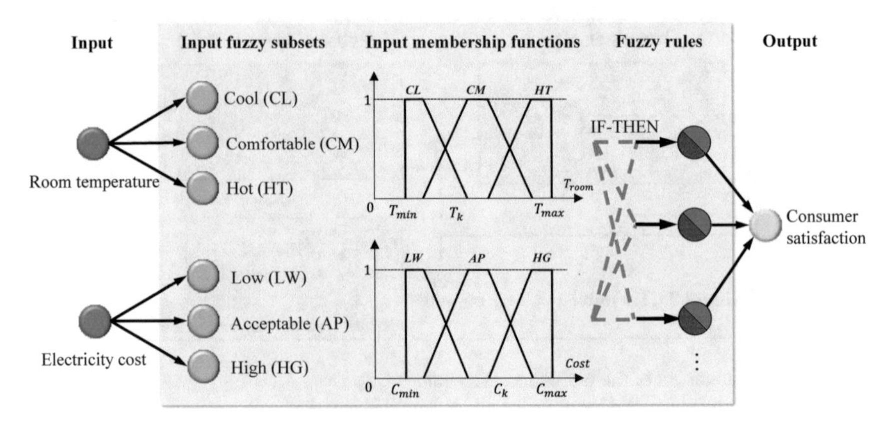

Fig. 3.2 The framework of fuzzy set method

$$T_{room}(t) = T_{room}(t-\tau) + \frac{Q - H_r(T_{room}(t-\tau) - T_{ext})}{C_{room}} \tag{3.5}$$

where H_r is the equivalent thermal conductance between the indoor air and the ambient air; T_{ext} and C_{room} are the ambient temperature and thermal mass of the room, respectively.

3.2.2 Consumer Satisfaction Quantization

It is crucial to ensure consumer satisfaction when controlling electricity consumption of TCLs so that consumers are willing to participant in DR programs. TCL consumer satisfaction is mainly affected by their preferences for the room temperature and the electricity cost, which tend to be unspecific and may change over time with high uncertainty. The fuzzy set method is able to model the qualitative aspects of human knowledge without precise quantitative analysis [30, 31]. Therefore, a typical fuzzy model (i.e., TSK (Takagi-Sugeno-Kang) fuzzy model [32]), which is adapted at processing intermediate values just like human cognition, is proposed to quantify consumer satisfaction, as shown in Fig. 3.2.

Consumer's cognitions can be modeled by the approach of fuzzy set method. Three fuzzy subsets of room temperatures T_{room} are defined to represent different feelings, including cool (CL), comfortable (CM) and hot (HT). Similarly, three fuzzy subsets of electricity costs including low (LW), acceptable (AP) and high (HG) are defined to express the consumer sensitivity. The membership values of each fuzzy subset could be derived from adopted membership functions.

Each combination of the above subsets will have corresponding consumer satisfaction values. Consumer satisfaction value depends on the transition from the temperature and cost subsets. In TSK fuzzy model, the transition is defined as a form of IF-THEN rules with a linear function integrating the room temperature and the electricity cost. A typical fuzzy rule to calculate satisfaction can be expressed as

$$\text{If } (C \text{ is LW}) \text{ and } (T_{room} \text{ is CM}), \text{ then } \left(y_s^i = f^i(C, T_{room})\right) \tag{3.6}$$

where y_s^i is the consumer satisfaction value of the ith fuzzy rule. $f^i(\cdot)$ is the linear function of the ith fuzzy rule, which can be expressed as

$$f^i(C, T_{room}) = a_0^i + a_1^i \cdot C + a_2^i \cdot T_{room} \tag{3.7}$$

where a_0^i, a_1^i, a_2^i are parameters for the ith consumer. Consumers' trade-off between room temperature and electricity cost can be simulated based on reasonable choices of these parameters, whose example is shown in Sect. 3.4.1.

From Eqs. (3.6) and (3.7), the satisfaction index can be calculated by the function $f^i(\cdot)$, if the electricity cost C is LW and the room temperature T_{room} is CM, respectively. Similarly, all combinations of the subsets can be calculated according to different parameters of heterogeneous customers. In this manner, consumer satisfaction y_s can be expressed as the output of TSK fuzzy model [32]

$$y_s = \frac{\sum_{i=1}^{R} (\mu_C^i(C) \cdot \mu_T^i(T_{room}) \cdot y_s^i)}{\sum_{i=1}^{R} (\mu_C^i(C) \cdot \mu_T^i(T_{room}))} \tag{3.8}$$

where μ_C^i, μ_T^i are the membership functions of the different fuzzy subsets corresponding to the ith fuzzy rule, whose subscript 'C' and 'T' indicate the fuzzy subsets of the electricity cost C and the room temperature T_{room}, respectively. R is the number of the fuzzy rules. From Eqs. (3.6) to (3.8), the TSK fuzzy model is able to output the quantified consumer satisfaction based on the inputs of the room temperature and the electricity cost.

3.2.3 *Maximum Satisfaction Control Strategies*

According to the economic man hypothesis [33], it is assumed that each individual is rational with complete knowledge and aims to maximize personal utility. Therefore, the decision making process of TCL consumer is to maximize the satisfaction, which is determined by the electricity cost and the room temperature. In this way, the control strategy of the set temperature is to maximize satisfaction y_s and is expressed by

$$Max \frac{\sum_{i=1}^{R} \left(\mu_C^i(C) \cdot \mu_T^i(T_{room}) \cdot y_s^i\right)}{\sum_{i=1}^{R} \left(\mu_C^i(C) \cdot \mu_T^i(T_{room})\right)} \tag{3.9}$$

where y_s is the consumer satisfaction expression shown in Eq. (3.8).

The constraints are as following:

(a) TCL model in Eqs. (3.1) and (3.2);
(b) Cost function in Eq. (3.3);
(c) Thermal model in Eqs. (3.4) and (3.5);
(d) Inequality constraint:

$$T_{\min} \le T_{set}(t) \le T_{\max} \tag{3.10}$$

The objective function of the Eq. (3.10) is complex and nonlinear, which lead to a nonlinear mixed-integer programming. In practice, the set temperature of a TCL is an integer and limited in a certain range. For example, the set temperature range of TCL is between 18 and 30 °C in general, where only the integer temperatures could be set. There are few temperature alternatives for consumer decisions. Therefore, traversal method is applied to solve this optimization. The calculation steps are as following:

(a) List alternative set temperatures of the TCL;
(b) Calculate the average power and electricity cost in different set temperatures;
(c) Calculate the consumer satisfaction values by the fuzzy model;
d) Compare each satisfaction values and choose the set temperature T_{set} corresponding to maximum satisfaction as the optimal decision.

As shown in Fig. 3.1, the set temperature T_{set} serves as an intermediate variable in the process from input (price signals) to output (power consumption). After obtaining optimal T_{set}, power consumption P can be calculated based on the electric-thermal model and then, ORC can be evaluated.

3.3 ORC Evaluation of Aggregated Heterogeneous TCLs

The individual TCL model involves many parameters and variables, which could influence the output more or less. Aggregators or operators of power systems are required to obtain all the values to evaluate the total ORC. In general, ORC evaluation can be calculated in

$$P_{ORC}(p_0, p_1) = \sum_{k=1}^{N} \Delta P^k(p_0, p_1) \tag{3.11}$$

where N is the total number of TCLs, and $\Delta P^k(p_0, p_1)$ is a function of the current price p_0 and target price p_1, which can be expressed as

$$\Delta P^k(p_0, p_1) = P^k(p_0) - P^k(p_1) \tag{3.12}$$

where $P^k(p)$ is the average power of the k-th TCL.

In practice, estimation sometimes must be made without enough information due to limited measurements or equipment failures. For example, the thermal parameters (e.g. equivalent thermal conductance and thermal mass) cannot be collected easily because the values vary with time, locations and buildings. These parameters or variables may be obtained via survey by field works, or by parameter fitting from actual monitoring data. But it costs too much to obtain these parameters of every individual, especially for a large-scale heterogeneous aggregation. Thus, ORC provided by TCLs must be evaluated based on insufficient data. In this section, two feasible estimation methods are given: moment estimation (ME) method and probability density estimation (PDE) method.

3.3.1 Moment Estimation Method

Moment estimation (ME) method, as one of the point estimation methods [34], is applied in ORC evaluation. The evaluation steps are shown in Fig. 3.3.

It is assumed that the number of known TCLs is N_s of the whole number N. The mean value of ORC based on N_s TCL data is calculated. Then, the total ORC of all the N TCLs is estimated, which can be expressed as

$$\widehat{P}_{ORC}(p_0, p_1) = \frac{N}{N_s} \sum_{k_s=1}^{N_s} \Delta P^k(p_0, p_1) \tag{3.13}$$

3.3.2 Probability Density Estimation Method

The probability density estimation (PDE) method is divided into two stages. Kernel density estimation is used in the first stage to estimate probability density function based on the limited measurement data. In the second stage, the ORC of aggregated TCLs is evaluated by calculating the expectation of power consumption.

(1) **The First Stage**

Kernel density estimation is one of non-parametric PDE methods to estimate the probability density distribution. Compared with parametric methods, the main advantage is the extensive applicability in unknown densities, especially for irregular shapes [35]. Moreover, the smoothness of kernel density results helps to avoid statistical errors compared with other non-parametric density estimation methods, such as frequency histogram. In Fig. 3.4, a multi-peak density function is taken as an example to illustrate the principle of kernel density estimation.

There are 6 known data points marked in black solid lines. Every known data point corresponds to a kernel function, which is normal distribution indicated by red

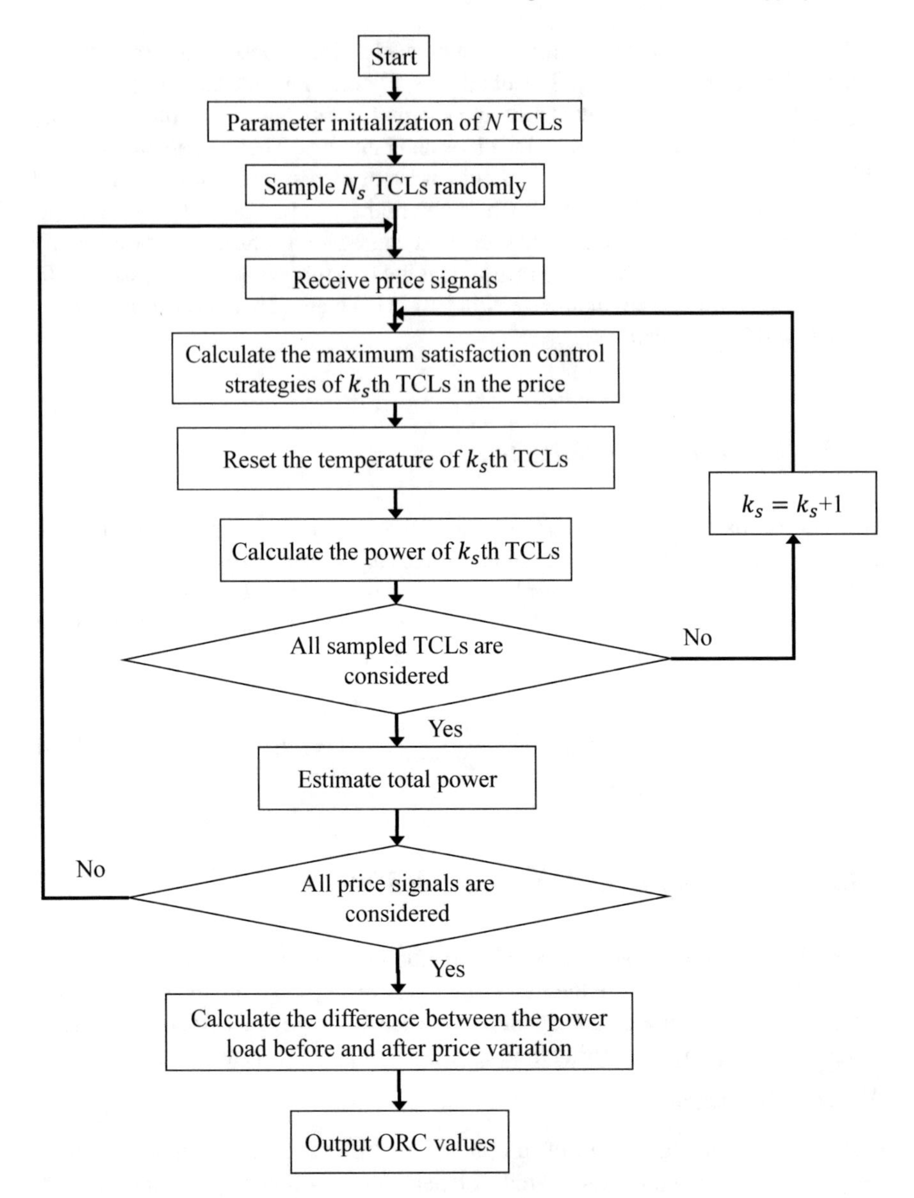

Fig. 3.3 The flow chart of ME method

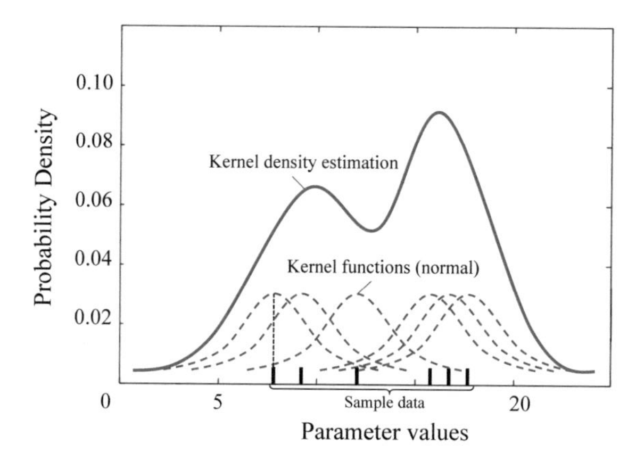

Fig. 3.4 The principle of kernel density estimation

dashed curves. These kernel functions are summed for the kernel density estimation, which is shown as the blue solid line.

In this way, the density of a parameter could be estimated based on the data with the least prior knowledge. To address this process, let $(H_r^1, H_r^2, \ldots H_r^{N_s})$ be independent and identically distributed variables of equivalent thermal conductance H_r, where N_s is the known data number in the whole number N. The estimated probability density $\hat{f}_{h_{Hr}}$ can be expressed as

$$\hat{f}_{h_{Hr}}(H_r) = \frac{1}{N_s h_{Hr}} \sum_{i=1}^{N_s} K\left(\frac{H_r - H_r^i}{h_{Hr}}\right) \tag{3.14}$$

where $K(\cdot)$ is the kernel function, which is nonnegative and integral value is 1. There are several kernel functions can be used, such as uniform, normal (Gaussian), Epanechnikov and others [35]. Considering the convenient mathematical properties, normal kernel function is applied:

$$K(x) = \frac{1}{(2\pi)^{n/2}} \exp\left(-\frac{x^2}{2}\right) \tag{3.15}$$

The h_{Hr} is the bandwidth of the kernel $K(\cdot)$, which is required to be chosen strictly for the trade-off between the deviation of the estimator and its variance. If the bandwidth is large, the results will be too smooth and omit some important information. On the contrary, the results will contain lots of noises if the bandwidth is small. Different bandwidth choices will lead to completely different estimation results. To minimize the mean integrated squared error [36], the rule-of-thumb bandwidth estimator method is adopted for normal kernel [37]

$$h_{Hr} = \left(4/3N_s\right)^{1/5} \hat{\sigma}_{Hr} \tag{3.16}$$

where $\hat{\sigma}_{Hr}$ is the standard deviation of H_r.

Similarly, the probability density function $\hat{f}_{hC_{room}}$ of the thermal mass of the room C_{room} can be obtained based on the kernel density estimation.

In order to evaluate the mean thermal potential, the joint density function of H_r and C_{room} should be derived from the marginal density functions of $\hat{f}_{hC_{room}}$ and $\hat{f}_{hH_r}$. H_r is mainly influenced by the room area and materials of walls, whereas C_{room} primarily depends on the room space and the air heat capacity. Hence, there is no direct relationship between H_r and C_{room}, which indicates the independence of the two parameters. The joint density function $\hat{f}_h$ can be expressed as

$$\hat{f}_h(C_{room}, H_r) = \hat{f}_{hC_{room}}(C_{room}) \cdot \hat{f}_{hH_r}(H_r) \tag{3.17}$$

From Eqs. (3.14) to (3.17), the joint density function can also be extended to multi-dimension if the parameters are independent with each other.

(2) **The Second Stage**

Mean value of individual power consumption can be obtained by calculating the probability expectation, which is described as

$$\widehat{P}_{avg}(p_0) = \int_{C_{room}} \int_{H_r} \hat{f}_h(C_{room}, H_r) \cdot P_{avg}(p_0, C_{room}, H_r) \tag{3.18}$$

where P_{avg} is the expectation of TCL power at the price p_0; x, y represent estimated parameters.

The ORC of the TCL can be evaluated as

$$\widehat{P}_{ORC}(p_0, p_1) = N \cdot (\widehat{P}_{avg}(p_0) - \widehat{P}_{avg}(p_1)) \tag{3.19}$$

The total ORC provided by the aggregated heterogeneous TCLs can be calculated as the flow chart in Fig. 3.5.

3.4 Case Studies

This section proves the efficiency of the proposed model and methods by case studies. Section 3.4.1 introduces the test system. Section 3.4.2 analyzes the accuracy of the ME method and the PDE method. Section 3.4.3 represents a real application of ORC evaluation in an actual case study.

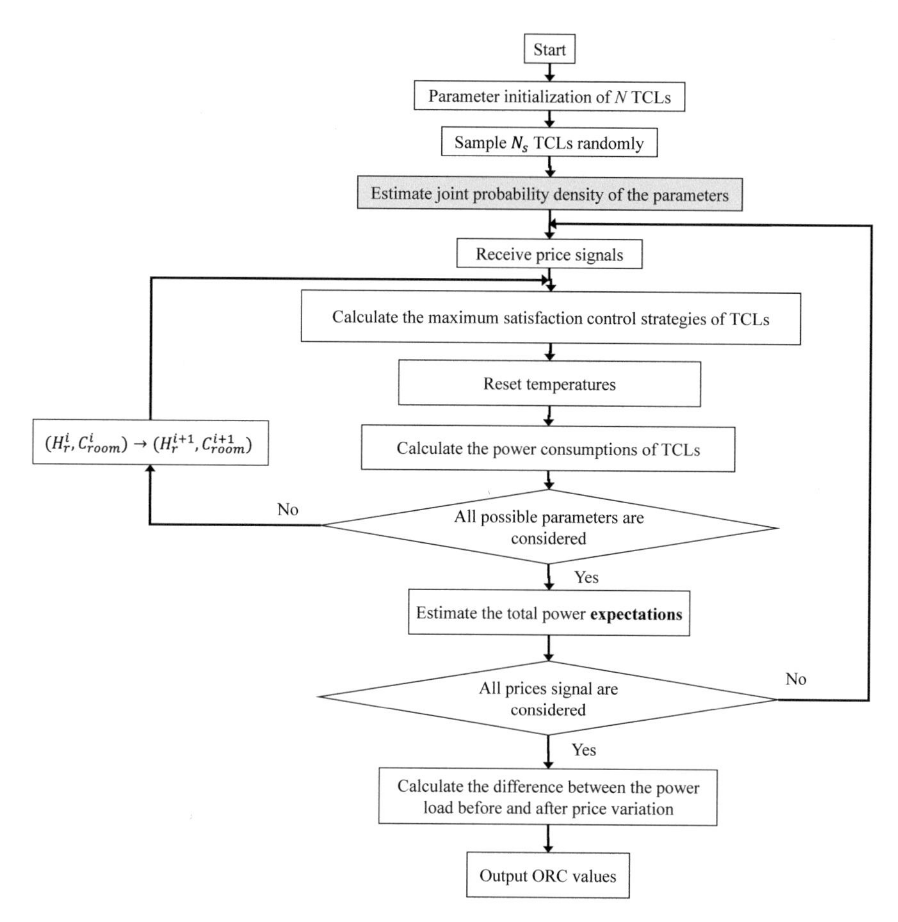

Fig. 3.5 The flow chart of PDE method

3.4.1 *The Test System*

In the test system, the ambient temperature is 30 °C. The total number of TCLs is 20,000, where only 100 TCLs can be randomly measured. The rated power of TCL is assumed to be 2 kW. The energy efficiency radio (EER) is 3.0. ΔT_0 is set to be 1 °C. It is assumed that the thermal mass of the room C_{room}(kJ/°C) obeys normal distribution, which is $C_{room} \sim N(12,\ 3.6^2)$. In order to generalize the distribution of equivalent thermal conductance H_r(kW/°C), half of H_r follows $N(1.5,\ 0.4^2)$ and the other half follows $N(0.8,\ 0.2^2)$ [3].

The initialized fuzzy subsets of the room temperature and the electricity cost are shown in Table 3.1 and Table 3.2, respectively [32]. Figure 3.6 shows the example of fuzzy subset.

Table 3.1 Fuzzy subsets of the room temperature

Fuzzy subsets	Min	Up-Min	Up-Max	Max
CL	–	–	21.0	24.0
CM	21.0	24.0	26.0	29.0
HT	26.0	29.0	–	–

Table 3.2 Fuzzy subsets of the electricity cost

Fuzzy subsets	Min	Up-Min	Up-Max	Max
LW	–	–	1.0	2.0
AP	1.0	2.0	4.0	5.0
HG	4.0	5.0	–	–

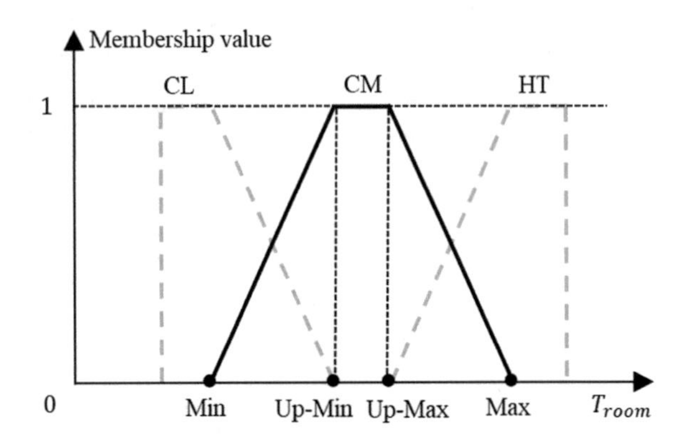

Fig. 3.6 Fuzzy subset example of CM (the comfortable feeling of the room temperature)

Table 3.3 Fuzzy rules of individual TCL model

$f^i(\cdot)$	CL	CM	HT
LW	$1-\alpha$	1	$1-\alpha$
AP	$\alpha\cdot(1-\alpha)$	α	$\alpha\cdot(1-\alpha)$
HG	$\alpha^2\cdot(1-\alpha)$	α^2	$\alpha^2\cdot(1-\alpha)$

In addition, the fuzzy rules are shown in Table 3.3. The parameter α is the weight of satisfaction to the different factors, indicating the consumer's trade-off between room temperature and electricity cost. The weight of room temperature is heavier with a larger α. All the α are assumed to follow uniform distribution $\alpha \sim U\ (0, 1)$.

Three cases are considered to illustrate the efficiency of proposed methods:

Case 1: Direct summation based on sufficient parameters of 20,000 TCLs. This case can be regard as actual value.

Case 2: Moment estimation (ME) method based on insufficient parameters of 100 TCLs.

Case 3: Probability density estimation (PDE) method based on insufficient parameters of 100 TCLs.

To compare this the estimation performance of the two method, the error of the estimated ORC is defined as

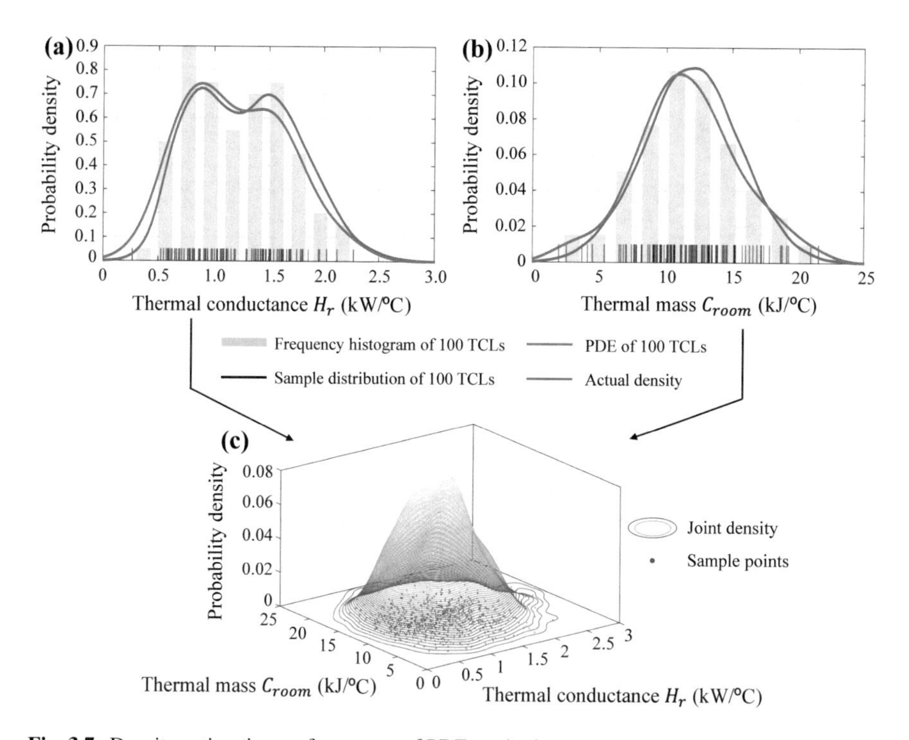

Fig. 3.7 Density estimation performances of PDE method

$$e_m = \left| \widehat{P}^{m}_{ORC} - \widehat{P}^{1}_{ORC} \right| \Big/ \widehat{P}^{1}_{ORC} \, (m = 2,\ 3) \tag{3.20}$$

where $\widehat{P}^{m}_{ORC}$ is the evaluated ORC of the aggregated TCLs in Case m.

3.4.2 ORC Evaluation with Insufficient Data

The first stage of the PDE method is to estimate the distribution of heterogeneous insufficient parameters by kernel density estimation. The accuracy of the proposed method is analyzed as shown in Fig. 3.7.

Figure 3.7a and b show the probability density distributions of the thermal mass C_{room} and the thermal conductance H_r, respectively. Both known data points are marked with a number of tiny black bars scattered on the axis of the abscissa. Frequency histograms and kernel density are marked with the gray histograms and the red curves, respectively. As shown in the figures, frequency histograms of 100 TCLs can reflect the general trend of the actual density (the blue curve). But there are obvious deviations due to abnormal data, such as the third histogram in Fig. 3.7a.

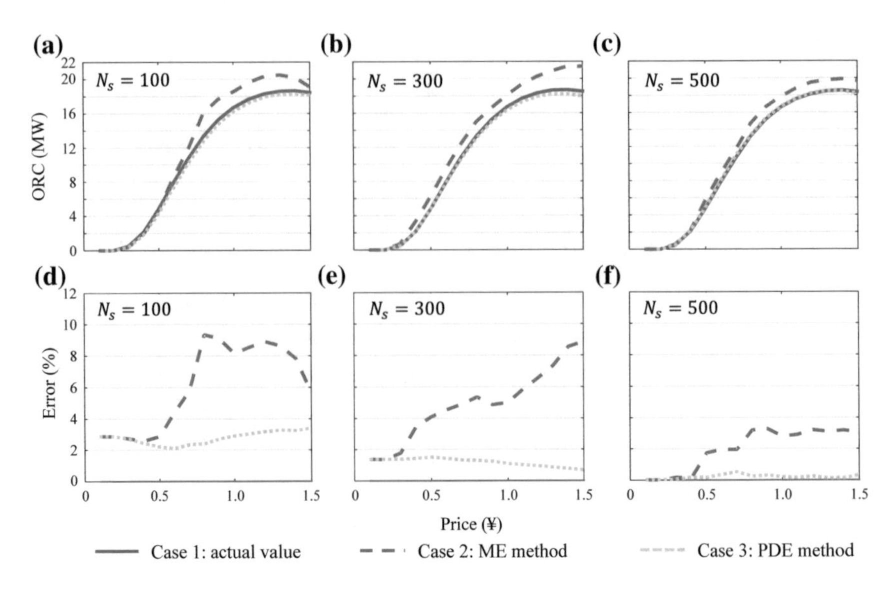

Fig. 3.8 ORC evaluation performances of the two methods

By contrast, probability density of 100 TCLs estimated by PDE method is almost overlapped with the actual density distributions, which is able to reduce the impact of the abnormal data and smooth the curve. The joint density distribution is obtained from the probability densities of H_r and C_{room}, shown in Fig. 3.7c.

Figure 3.8 shows the ORC evaluation performances of different methods in different number size N_s of known data. When $N_s = 100$, compared with the curve of ME method (the red dotted curves), the curve of PDE method (the orange dotted curves) is much closer to the actual curve (the blue solid curves). The error comparison is shown in Fig. 3.8d, where the errors reach over 9.0% in ME method, whereas the errors in PDE method are within 4.0%. Similar conclusions can be made in other values of N_s. Therefore, the proposed PDE method is more accurate than ME method, and is able to improve the ORC evaluation accuracy based on the same insufficient data.

Figure 3.8d, e and f show the trend of evaluation errors of ME method and PDE method in different N_s, respectively. In comparison of the three figures, both errors of ME method and PDE method experience a significant decrease with the increase of N_s. More details of evaluation errors are explained in Fig. 3.9.

In Fig. 3.9, average errors corresponding to different number size N_s are calculated to highlight the PDE method applicability in different data distributions. With the increase of N_s, the average errors of both methods decline and converge to zero gradually. In comparison, the average error of PDE method is significantly less than that of ME method. The error differences of the two methods are shown in Fig. 3.8 (the black solid curve), where the maximum difference reaches 7.8% at the number size 50. The trend of curve shows that there are less error differences in larger number

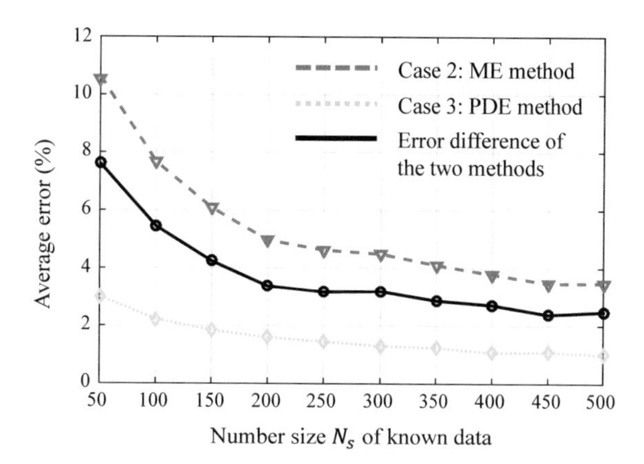

Fig. 3.9 Average error of ORC evaluation in different number size

sizes, which highlights the estimation accuracy of PDE method especially in small number size of known data.

3.4.3 ORC Evaluation in Actual Case Studies

In this subsection, the practicality of the proposed PDE method is verified in the case studies based on actual DR data. Firstly, one of the pilot projects in a province of China is introduced with actual data. Then it is illustrated that the proposed PDE method can be applied in possible events of the pilot project. Finally, case study and analysis of PDE method applications are discussed.

(1) **Introduction of the Pilot Project**

One of the pilot projects selects 522 residential consumers in Jiangsu province of China, where the power consumption of TCLs accounts for more than 30% during summer peak load. Smart meters and terminal controllers are installed in order to enable consumers to make demand response strategies with their personal needs.

The aggregated power consumption of selected consumers is collected in every 15 min for two weeks under similar weather conditions. The first week data without demand response (DR) program is regarded as baseline load. During the second week, the peak price signals were sent to consumers between 14:00 and 15:00 every day and thereby, the power consumption decreased to provide ORC for power system. The power consumptions are averaged based on the obtained data of the two week.

Results of actual DR program are shown in Fig. 3.10a, where the blue solid curve is the sum of power in non-DR case, while the black solid curve is the sum of power in DR case. These two curves are overlapped at most of the time except for the period between 14:00 and 16:00, where the DR case experienced a significant load curtailment and the power consumption decreased to the minimum at 14:56. The operating reserve capacity (ORC) is 1.22 MW, calculated by the maximum load

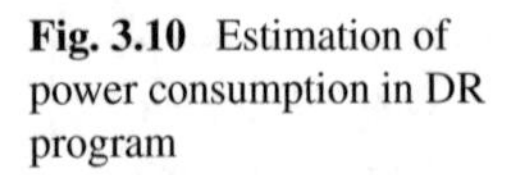

Fig. 3.10 Estimation of power consumption in DR program

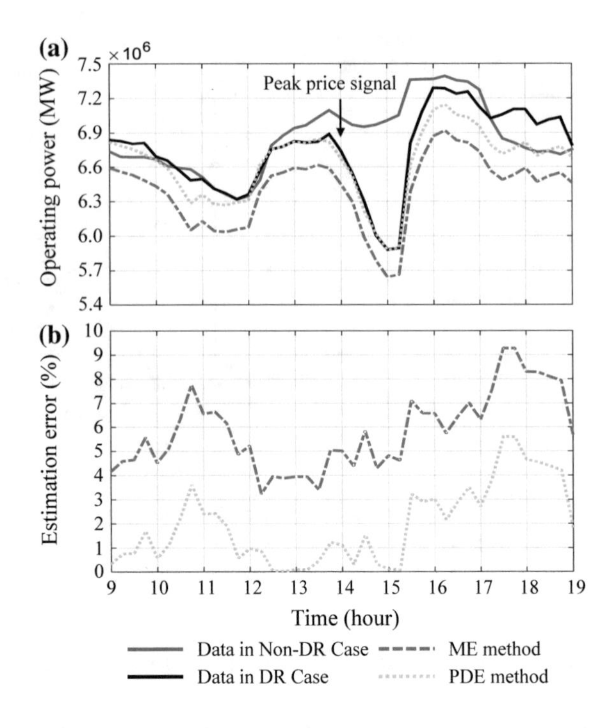

curtailment. It proves the feasibility of demand response providing operating reserve for power systems.

(2) Case Studies of Proposed Method Applications

The proposed PDE method, which provides a more accurate estimation with less measured data, can be widely applied in demand response programs. For example, in the case of data loss due to communication or measurement failure, complete data of every individual cannot be obtained to calculate aggregated power consumption, which may serve as an important index for further actions. In the circumstances, the proposed PDE method provides an approach to improve the accuracy of estimation with limited available data, which could help to make the right decisions.

The following case study shows the application of PDE method in the above pilot project when there exists data loss. The total number of residential consumers is N in the pilot project. It is assumed that only the data of N_s consumers can be obtained while other data is lost due to equipment failures. The ME method and PDE method are applied to estimate the power consumption with N_s known consumers, respectively. N and N_s are set to be 522 and 50, respectively. The flow chart of this case studies is shown in Fig. 3.11.

Figure 3.10a shows the estimation of power consumption with ME method and PDE method, which is illustrated by the red dotted curve and the orange dotted curve, respectively. The estimation of power consumption obtained by PDE method is much more consistent with that of ME method. The error between estimation data

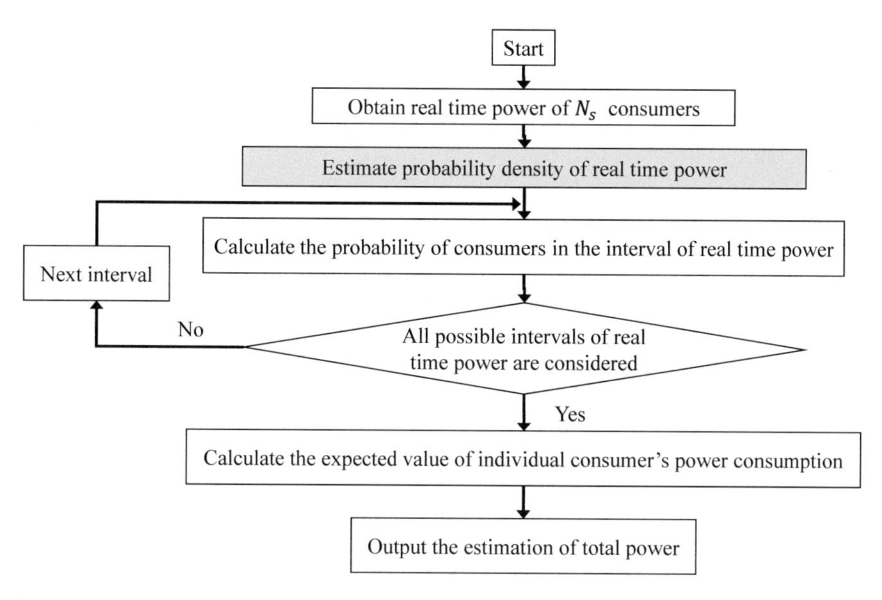

Fig. 3.11 The flow chart of estimation of power consumption

and actual data in DR program (the black solid curve) is shown in Fig. 3.10b, where PDE method is more accurate and appropriate than ME method. The average error of ME method reaches 5.90%, which is relatively high value. Compared to that, the average error of PDE method is 2.52%, which is much less than that of ME method and thereby, proves the practicality of PDE method to estimate aggregated power with insufficient data.

3.5 Conclusions

The progress of information and communication technology has made it easier for demand side resources to provide operating reserve. This chapter proposes a quantitative evaluation method of operating reserve capacity (ORC) provided by aggregated heterogeneous TCLs with insufficient measurement data. The individual TCL model considering consumer's behaviors is developed to characterize the impact of fluctuated electricity prices and different temperature requirements. Consumer's perspective on electricity prices and feeling of the room temperature are modeled by the fuzzy set method. In this manner, the consumer's satisfaction is defined and then optimized in the maximum satisfaction control strategies. Moreover, the probability density estimation (PDE) method is proposed to evaluate the ORC provided by large-scale heterogeneous TCLs without enough measurement data. The numerical studies show that, compared with the traditional estimation method, the PDE method can improve the accuracy of the ORC evaluation with insufficient data.

References

1. O. Ma, N. Alkadi, P. Cappers, P. Denholm, J. Dudley, S. Goli, M. Hummon, S. Kiliccote, J. MacDonald, N. Matson, D. Olsen, C. Rose, M.D. Sohn, M. Starke, B. Kirby, M. O'Malley, Demand response for ancillary services. IEEE Trans. Smart Grid **4**(4), 1988–1995 (2013)
2. J. Wang, X. Wang, Y. Wu, Operating reserve model in the power market. IEEE Trans. Power Syst. **20**(1), 223–229 (2005)
3. H. Hui, Y. Ding, D. Liu, Y. Lin, Y. Song, Operating reserve evaluation of aggregated air conditioners. Appl. Energy **196**, 218–228 (2017)
4. Z. Bricman Rejc, M. Cepin, Estimating the additional operating reserve in power systems with installed renewable energy sources. Int. J. Electr. Power Energy Syst. **62**, 654–664 (2014)
5. P. Siano, D. Sarno, Assessing the benefits of residential demand response in a real time distribution energy market. Appl. Energy **161**, 533–551 (2016)
6. J. Hu, J. Cao, Z.Q.M. Chen, J. Yu, J. Yao, S. Yang, T. Yong, Load following of multiple heterogeneous TCL aggregators by centralized control. IEEE Trans. Power Syst. **32**(4), 3157–3167
7. W. Zhang, J. Lian, C.Y. Chang, K. Kalsi, Aggregated modeling and control of air conditioning loads for demand response. IEEE Trans. Power Syst. **28**(4), 4655–4664 (2013)
8. J.C. Hwang, Assessment of air condition load management by load survey in Taipower. IEEE Trans. Power Syst. **16**(4), 910–915 (2001)
9. K. Wang, J. Yao, L. Yao, S. Yang, T. Yong, Survey of research on flexible loads scheduling technologies. Autom. Electr. Power Syst. **38**(20), 127–136 (2014)
10. D.S. Callaway, Tapping the energy storage potential in electric loads to deliver load following and regulation with application to wind energy. Energy Convers. Manag. **50**(5), 1389–1400 (2009)
11. D. Papadaskalopoulos, G. Strbac, P. Mancarella, M. Aunedi, V. Stanojevic, Decentralized participation of flexible demand in electricity markets—Part II: application with electric vehicles and heat pump systems. IEEE Trans. Power Syst. **28**(4), 3667–3674 (2013)
12. K. Dehghanpour, S. Afsharnia, Electrical demand side contribution to frequency control in power systems: a review on technical aspects. Renew. Sustain. Energy Rev. **41**, 1267–1276 (2015)
13. N. Lu, Y. Zhang, Design considerations of a centralized load controller using thermostatically controlled appliances for continuous regulation reserves. IEEE Trans. Smart Grid **4**(2), 914–921 (2013)
14. M.L. Zheng, R.Y. Fang, Z.T. Yu, Life cycle assessment of residential heating systems: a comparison of distributed and centralized systems, in *Applied Energy Symposium and Forum 2016: Low Carbon Cities & Urban Energy Systems* (2016)
15. F. Luo, Z.Y. Dong, K. Meng, J. Wen, H. Wang, J. Zhao, An operational planning framework for large-scale thermostatically controlled load dispatch. IEEE Trans. Ind. Inform. **13**(1), 217–227 (2017)
16. N. Lu, An evaluation of the HVAC load potential for providing load balancing service. IEEE Trans. Smart Grid **3**(3), 1263–1270 (2012)
17. G. Bianchini, M. Casini, A. Vicino, D. Zarrilli, Demand-response in building heating systems: a model predictive control approach. Appl. Energy **168**, 159–170 (2016)
18. M. Liu, Y. Shi, Model predictive control of aggregated heterogeneous second-order thermostatically controlled loads for ancillary services. IEEE Trans. Power Syst. **31**(3), 1963–1971 (2016)
19. X. He, N. Keyaerts, I. Azevedo, L. Meeus, L. Hancher, J.M. Glachant, How to engage consumers in demand response: a contract perspective. Util. Policy **27**, 108–122 (2013)
20. D. Qv, B.B. Dong, L. Cao, L. Ni, J.J. Wang, R.X. Shang, Y. Yao, An experimental and theoretical study on an injection-assisted air-conditioner using R32 in the refrigeration cycle. Appl. Energy **185**, 791–804 (2017)
21. R. Yu, W. Yang, S. Rahardja, A statistical demand-price model with its application in optimal real-time price. IEEE Trans. Smart Grid **3**(4), 1734–1742 (2012)

22. D.P. Chassin, D. Rondeau, Aggregate modeling of fast-acting demand response and control under real-time pricing. Appl. Energy **181**, 288–298 (2016)
23. N. Alibabaei, A.S. Fung, K. Raahemifar, A. Moghimi, Effects of intelligent strategy planning models on residential HVAC system energy demand and cost during the heating and cooling seasons. Appl. Energy **185**, 29–43 (2016)
24. C. Gu, X. Yan, Z. Yan, F. Li, Dynamic pricing for responsive demand to increase distribution network efficiency. Appl. Energy **205**, 236–243 (2017)
25. H. Allcott, Rethinking real-time electricity pricing. Resour. Energy Econ. **33**(4), 820–842 (2011)
26. J.A. Gomez, M.F. Anjos, Power capacity profile estimation for building heating and cooling in demand-side management. Appl. Energy **191**, 492–501 (2017)
27. D. Xie, H. Hui, Y. Ding, Z. Lin, Operating reserve capacity evaluation of aggregated heterogeneous TCLs with price signals. Appl. Energy **216**, 338–347 (2018)
28. H. Chao, Price-responsive demand management for a smart grid world. Electr. J. **23**(1), 7–20 (2010)
29. S. Ihara, F.C. Schweppe, Physically based modeling of cold load pickup. IEEE Trans. Power App. Syst. **100**(9), 4142–4150 (1981)
30. K. Bhattacharyya, M.L. Crow, A fuzzy logic based approach to direct load control. IEEE Trans. Power Syst. **11**(2), 708–714 (1996)
31. J.-S.R. Jang, ANFIS: adaptive-network-based fuzzy inference system. IEEE Trans. Syst. Man Cybern. **23**(3), 655–685 (1993)
32. T. Takagi, M. Sugeno, Fuzzy identification of systems and its applications to modeling and control. IEEE Trans. Syst. Man Cybern. **15**(1), 116–132 (1985)
33. J. Persky, Retrospectives: the ethology of homo economicus. J. Econ. Perspect. **9**(2), 221–231 (1995)
34. J.M. Morales, Point estimate schemes to solve the probabilistic power flow. Reliab. Eng. Syst. Saf. **22**(4), 1594–1601 (2007)
35. V.A. Epanechnikov, Non-parametric estimation of a multivariate probability density. Theory Probab. Appl. **14**, 153–158 (1969)
36. A. Molina-García, M. Kessler, J.A. Fuentes, E. Gómez-Lázaro, Probabilistic characterization of thermostatically controlled loads to model the impact of demand response programs. IEEE Trans. Power Syst. **26**(1), 241–251 (2011)
37. A.W. Bowman, A. Azzalini, *Applied Smoothing Techniques for Data Analysis* (Oxford University Press Inc., New York, 1997)

Chapter 4
Air Conditioner Aggregation for Providing Operating Reserve Considering Lead-Lag Rebound Effect

4.1 Introduction

The above chapters show that the growing penetration of renewable energy sources into the electric power system calls for a huge amount of balancing services at multiple timescales [1, 2]. Air conditioners (ACs) offer an alternative of traditional generation units for balancing the system by actively reducing or increasing electricity consumption [3–6]. With the development of smart grid technologies and real-time telemetry [7, 8], it is technically feasible for ACs to respond to instructions within a short period and provide operating reserve at various time scales [9]. However, demand response rebound is one possible obstacle of using ACs for the provision of operating reserve [10]. This phenomenon is the rebound peak that arises when a large amount of loads are re-connected to the grid at approximately the same time [11]. The existence of the demand response rebound may cause significantly higher demand than that prior to the demand response event. In extreme cases, the increased load current derived from the rebound peak may even lead to the melting of overhead lines, which harms system security considerably [12].

The demand response rebound at the end of a demand response event, which is when ACs are recovered to the initial states, is also referred to as load payback effect [13, 14], load recovery effect [15, 16] or cold load pickup [12, 17] in the existing literatures. This phenomenon has been observed in many pilot projects, including the Californian pilot study of time-of-use and critical peak pricing [18] and the Norwegian project of direct control of residential water heaters in 475 households [19]. Several studies have addressed the impacts of demand response rebound on the market scheme [20, 21] and system scheduling [10, 22]. Most researches reduce the level of rebound by increasing the diversity of loads [19] or randomizing the reconnection of appliances over time [13]. Apart from that, reference [23] copes with demand response rebound by leveraging chilled-water capacity through a least enthalpy estimation based thermal comfort control. The above approaches can reduce the demand response rebound to some extent but cannot mitigate the rebound entirely. The concept of dispatching groups of devices in sequence is initially presented to prolong the

Y. Ding et al., *Integration of Air Conditioning and Heating into Modern Power Systems*,
https://doi.org/10.1007/978-981-13-6420-4_4

deployment duration of operating reserve without increasing the interruption duration of individual consumers [24, 25]. Reference [26] further indicates that restoring group of devices to the initial state in sequence can also reduce the demand response rebound. This is examined by the recovery of water heaters in [17, 27], which illustrate that both the magnitude and mean value of demand response rebound can be largely reduced. However, the devices are evenly divided into several groups and are recovered at a regular time interval (e.g., 10 min) in [17, 24, 25, 26, 27]. As a result, the aggregate dynamics of devices are not considered, which cannot guarantee that the rebound peak is reduced to the minimum value. Reference [13] presents the concept to achieve better control on the demand response rebound by the coordination among different groups of devices, but there lacks mathematical model to determine the coordination process among the groups. Moreover, these research studies only consider the rebound load during the recovery period, while neglecting the rebound load during the reserve deployment period.

For clarity, rebound peak of aggregate power during the reserve deployment period is named as the lead rebound effect in this chapter. By contrast, rebound peak of aggregate power during the recovery period is named as the lag rebound effect. The lead rebound effect is neglected in previous research studies because the ACs are converted to the *off* state when providing operating reserve. Hence, ACs can reliably provide load reduction with different duration time, as long as they remain in the *off* state [10, 22]. However, shutting the units off directly may cause short-cycling of ACs, which can reduce their lifetime, increase maintenance, and potentially damage them [28, 29]. Therefore, recent research studies have focused on controlling ACs through changing the set point temperature instead [11, 30]. In this case, according to the Law of Energy Conservation [23], demand response rebound will also occur during the reserve deployment period. For example, ACs are in the cooling mode in summer. Upon receiving the load-reduction instruction, ACs will migrate to the new upper temperature hysteresis band and stay longer in the *standby* state. More internal heat will accumulate with longer *standby* period of ACs. Hence, higher demand for cooling occurs when ACs have reached the new upper temperature hysteresis band, resulting in the increase of electricity consumption, i.e., lead rebound effect.

The lead rebound effect poses a new challenge to the provision of the operating reserve by ACs because it limits the duration time before the rebound occurs. If the rebound occurs within the period of reserve deployment, then ACs cannot sustain the required reserve capacity in the required duration time. Requirements on the duration time of the operating reserve are crucial because it ensures the system operator has adequate time to correct the imbalance between load and generation [31]. Typically, the required duration time of spinning/non-spinning reserve can be 30 min or even 60 min [32]. Considering that the lead rebound may occur approximately 10 min after the control of ACs, ACs encounter difficulty in fulfilling the requirements on the duration time. To make the most of the ACs' potential to provide operating reserve, the duration time of the load reduction or load increase should be flexibly controlled.

This entails a need to mitigate the lead rebound entirely during the required reserve deployment period, which can be potentially realized by dispatching different groups of ACs in sequence. However, as illustrated above, there lacks the consideration on the dynamics of ACs, and so does the co-optimization among the reserve capacity and dispatch time instant of different AC groups in existing literatures, making it difficult to guarantee the entire mitigation of the lead-lag rebound effect. Moreover, the control of the rebound time, which is crucial for the determination of the duration time, is not involved. Consequently, the time of rebound remains uncontrollable and the utilization of operating reserve provided by ACs is still limited by the rebound period.

This chapter proposes an optimal sequential dispatch strategy of ACs to mitigate the lead-lag rebound entirely and thus realize the flexible provision of various types of operating reserve. Because of the constraint of the duration time imposed by the lead rebound, traditional methods that quantify the duration time between the reserve deployment and the recovery are not suitable [32]. Therefore, the first step is to develop an evaluation framework of operating reserves to quantify the effect of the rebound load on the capacity dimension and time dimension. Then, ACs dispatched at the same time instant are defined as an AC group, and different AC groups are dispatched in sequence. In order to guarantee that the rebound load of each group is mitigated entirely by the reserve capacity of the latter group, the dispatch time instant of each group is optimized to minimize the deviation between the actual load changing and the required value, while the selection of ACs in each group are optimized to make full use of ACs' available reserve capacity and guarantee consumers' comfort. Co-optimization of the above problems on the time dimension and capacity dimension forms a mixed integer nonlinear bi-level programming problem, which is then solved by genetic algorithm. In addition, a three-layer structure is designed to integrate the proposed strategy with the interactions between aggregators and consumers. Case studies are conducted to verify the proposed strategy for providing operating reserve with multiple duration time. The major contributions of this chapter are as follows:

(a) The lead rebound effect, which results in the special obstacle of controlling the duration time, is considered for the provision of operating reserve with ACs. To the best of the authors' knowledge, it is the first time that the lead rebound is modeled and analyzed.
(b) A capacity-time evaluation framework of the operating reserve provided by ACs is developed to quantify the impacts of the lead-lag rebound effect. Compared with the existing evaluation method for traditional generating units [32], the proposed evaluation framework can better characterize the dynamics of demand-side operating reserve.
(c) An optimal sequential dispatch strategy, which can entirely mitigate both the lead rebound and lag rebound, is proposed to realize the flexible control of duration time from minutes to several hours.

This chapter includes research related to the evaluation and sequential-dispatch of operating reserve provided by ACs considering lead-lag rebound effect by [33].

4.2 Analysis of the Lead-Lag Rebound Effect

4.2.1 Model of an Individual AC

The operation process of an individual AC is described by the general state model for the thermostatically-controlled-loads (TCLs) [34]:

$$\frac{d\theta_i(t)}{dt} = -\frac{1}{C_i R_i}[\theta_i(t) - \theta_a(t) + m_i(t) R_i Q_i] \tag{4.1}$$

where $\theta_i(t)$ is the room temperature corresponding to the i-th AC at time t, $\theta_a(t)$ is the ambient temperature. C_i and R_i are the thermal capacity and thermal resistance corresponding to the room of the i-th AC, respectively. $m_i(t)$ represents the *on* or *standby* state of the i-th AC. Q_i is the energy transfer rate of the i-th AC, which is equal to the product of the input power p_i and the coefficient of performance COP_i of the i-th AC.

The i-th AC operates cyclically around its set point temperature $T_{set,i}$ with a dead band of ΔT_i. For example, if the AC is in the cooling mode in summer, it will switch to the *on* state when the room temperature reaches the upper band ($\theta_i^+ = T_{set,i} + 0.5 \times \Delta T_i$). Similarly, when the room temperature reaches the lower band ($\theta_i^- = T_{set,i} - 0.5 \times \Delta T_i$), it will switch to the *standby* state. The temperature range between θ_i^- and θ_i^+ is defined as the hysteresis band $[\theta_i^-, \theta_i^+]$ [34]:

$$m_i(t) = \begin{cases} 1, & \theta_i(t) > \theta_i^+ \\ 0, & \theta_i(t) < \theta_i^- \\ m_i(t-1), & otherwise \end{cases} \tag{4.2}$$

4.2.2 Aggregate Response of ACs

The AC load model (4.1)–(4.2) reveals that the state of ACs can be quickly controlled by changing the set point temperature. However, if ACs are controlled by consumers manually, the response may not be sufficiently fast to meet the requirement of ramp time [31]. Therefore, it is assumed that the ACs are installed with smart meters. By signing contracts with consumers, aggregators can control ACs at the permitted periods [35]. When the duration time has reached the required value, ACs are recovered to the initial states [13].

Denoting $\mathbf{\Gamma}$ as the set of all the ACs under an aggregator, the number of ACs in $\mathbf{\Gamma}$ is $N^{\max}$. The i-th AC is permitted to be controlled by the aggregator during the period $[t_i^s, t_i^e]$. The reserve deployment period instructed by the system operator is $[t_{ins}, t_{end}]$. The available AC is defined as the unit that is permitted to be controlled by the aggregator during the required duration. The availability of ACs is labeled by the vector $\mathbf{V} \in \mathbb{R}^{N^{\max} \times 1}$, in which the i-th element is:

$$v_i = \begin{cases} 1\ , & [t_{ins}, t_{end},] \in [t_i^s, t_i^e] \\ 0\ , & otherwise \end{cases}\ , \forall i \in \mathbf{\Gamma} \tag{4.3}$$

The ACs dispatched at the same time instant τ_g are defined as an AC group g. $\mathbf{S}_g \in \mathbb{R}^{N^{\max} \times 1}$ is the vector that represents the ACs belonging to group g. The i-th element in $\mathbf{S}_g$ is:

$$s_{g,i} = \begin{cases} 1\ , & i \in group\ g \\ 0\ , & otherwise \end{cases}\ , \forall i \in \mathbf{\Gamma} \tag{4.4}$$

The physical parameters of the i-th AC, including C_i and R_i, can be usually assumed as a constant value. The input power p_i is usually set by the AC manufacturer and cannot be changed by consumers. Considering that there may exist tens of thousands of units under an aggregator, it may be difficult to obtain the parameters of all the ACs. In this case, aggregators can randomly select N^e ACs to measure their parameters. Kernel density estimation, which is one of the non-parametric probability density estimation methods, can be utilized to obtain the probability density distribution of parameters from the known data points corresponding to N^e selected ACs [36]. For example, $\left(R_1^e,\ R_2^e,\ \ldots,\ R_{N^e}^e\right)$ denotes the thermal resistance of the randomly selected N^e ACs. The estimated probability density $\hat{f}_{R^e}$ of the thermal resistance R^e can be expressed as:

$$\hat{f}_{R^e}(R^e) = \frac{1}{N^e h_{R^e}} \sum_{i_e=1}^{N^e} K\left(\frac{R^e - R_{i_e}^e}{h_{R^e}}\right) \tag{4.5}$$

where $K(\cdot)$ is the normal kernel function. h_{R^e} is the bandwidth of the kernel function and can be determined by the rule-of-thumb bandwidth estimator method [36]. Similarly, the probability density of other parameters can also be estimated with the kernel density estimation. Then, the parameters of the ACs can be randomly set according to the probability density distribution of these parameters.

The temperature hysteresis band of individual AC in $\mathbf{\Gamma}$ is assembled in the vector $\boldsymbol{\theta}^{-(+)} = \left[\theta_1^{-(+)}, \theta_2^{-(+)} \cdots \theta_{N^{\max}}^{-(+)}\right]^{\mathrm{T}}$. Changes of the set point temperature during the reserve deployment/recovery of group g are assembled in the vector $\boldsymbol{\gamma}_g^{d/r} = \left[\gamma_{g,1}^{d/r}, \gamma_{g,2}^{d/r} \cdots \gamma_{g,N^{\max}}^{d/r}\right]^{\mathrm{T}}$. Similarly, $\bar{\boldsymbol{\gamma}}^{d/r}$ and $\underline{\boldsymbol{\gamma}}^{d/r}$ are the highest increase and decrease of the set point temperature determined by consumers, respectively. Apart from τ_g and the time t, the aggregate power is primarily influenced by param-

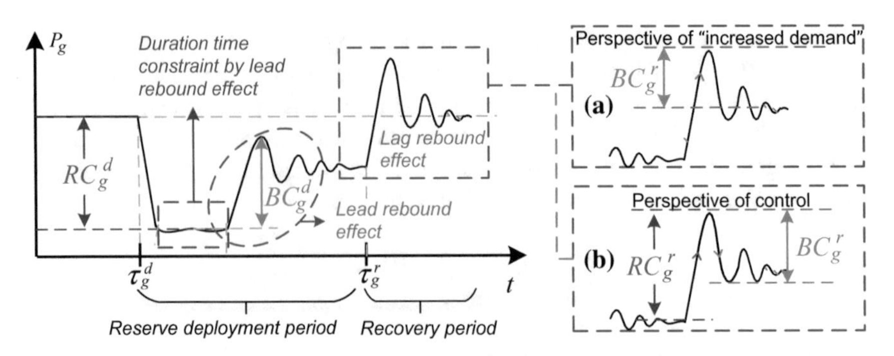

Fig. 4.1 Typical curve of the aggregate response of ACs

eters $\boldsymbol{\theta}^{-(+)}$, $\boldsymbol{\gamma}_g^{d/r}$ and $\mathbf{S}_g$, which are assembled in the array $\mathbf{u}_g = [\boldsymbol{\theta}^+, \boldsymbol{\theta}^-, \boldsymbol{\gamma}_g^{d/r}, \mathbf{S}_g] \in \mathbb{R}^{N^{\max}} \times 1$. Consequently, the aggregate power P_g of group g is a function of the mentioned parameters:

$$P_g(\mathbf{u}_g, \tau_g, t) = \sum_{i \in \boldsymbol{\Gamma}} p_i \cdot m_i(t) \cdot s_{g,i} \tag{4.6}$$

4.2.3 Lead Rebound Effect and Lag Rebound Effect

A typical curve of the aggregate response of an AC group is shown in Fig. 4.1, in which the ACs change the set point temperature for load reduction at τ_g^d and recover the set point temperature to the initial value at τ_g^r. Two rebound peaks of aggregate power exist during the reserve deployment period and the recovery period, respectively. In this chapter, the former is named as the lead rebound effect, and the latter is named as the lag rebound effect.

During the reserve deployment period, the decrease of aggregate power is corresponding to the provision of reserve capacity RC_g^d, while the increase of aggregate power is corresponding to the rebound capacity BC_g^d of the lead rebound. The lead rebound effect is rarely considered in existing research studies because it only occurs when ACs are controlled by changing the set point temperature. This property can be explained by the migration of ACs' room temperature through the Law of Energy Conservation [23]. For example, ACs are in cooling mode in summer. The variation of room temperature and the corresponding consumed power of the *i*-th AC are illustrated in Fig. 4.2. Upon receiving the load-reduction instruction at τ_g^d, the *i*-th AC increases the set point temperature and remains in the *standby* state until the room

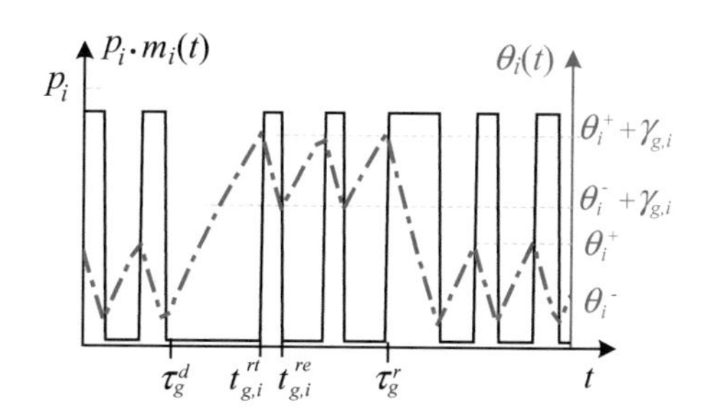

Fig. 4.2 Consumed power and room temperature of the i-th AC when the temperature hysteresis band is increased by $\gamma_{g,i}$ at τ_g^d and decreased by $\gamma_{g,i}$ at τ_g^r

temperature reaches the new upper temperature hysteresis band at $t_{g,i}^{rt}$, after which the AC will switch to the *on* state. The internal heat is not transferred outdoors and thus accumulates during the longer *standby* period shown by the segment from τ_g^d to $t_{g,i}^{rt}$. Hence, higher load demand occurs in the subsequent *on* state. Consequently, the lead rebound is the nature increase of aggregate power resulted from the cyclic operation characteristics of ACs. It is different from the opt-out behavior of consumers, which is the consumers' initiative behavior to choose not to participate in the demand response event when their comfort levels cannot be maintained [37, 38].

By contrast, the lag rebound effect is the result of consumers' recovery behavior and a common phenomenon after a demand response event [12, 19]. As shown in Fig. 4.2, the AC decreases the set point temperature after receiving the load-reduction instruction at τ_g^r and remains in the *on* state for longer time. Hence, aggregate power of ACs increases instantly when the loads are recovered to the initial states at τ_g^r. The lag rebound capacity BC_g^r is considered as the increase of aggregate power larger than the initial value in the existing research studies [12, 19], as illustrated in Fig. 4.1a. From the control perspective in this chapter, the recovery process can be regarded as the reserve deployment process with the required reserve capacity provided by the load-increase operation. In this case, the increase of aggregate power is corresponding to the provision of reserve capacity RC_g^r, while the decrease of aggregate power is corresponding to the rebound capacity BC_g^r, as illustrated in Fig. 4.1b.

As depicted in Fig. 4.1, the duration time is constrained within the period before the lead rebound occurs. It is required to notice that sudden changes of the set point temperature will cause temporary synchronization of the ACs [34]. In other words, after the ACs have reached the new temperature hysteresis band, it will switch between the *on* state and *standby* state at the approximately same time, which will increase the level of rebound and lead to large power fluctuations. However, few methods are available to quantify the effect of lead-lag rebound and the associated power fluctuations, making it difficult to mathematically describe the objectives of AC load control. Therefore, section III proposes several indices to evaluate operating reserves considering the effect of lead-lag rebound and power fluctuations.

4.3 Capacity-Time Evaluation of the Operating Reserve Considering Lead-Lag Rebound Effect

Operating reserve providers are required to respond to different types of events over different time frames [31]. Therefore, two fundamental dimensions, *capacity* and *time* [32], are required to be considered for evaluating the operating reserve. The capacity dimension entails assigned amount of load reduction/increase during the reserve deployment period. The time dimension involves duration time and ramp rate. The effect of the lead-lag rebound and the associated power fluctuations are also quantified on these two dimensions.

4.3.1 Universal Expression of the Load Reduction/Increase

The difference of the aggregate power before and after the changes of set point temperature represents the effects of AC load control. The aggregate power changes in an adverse direction in load-reduction and load-increase operation. Therefore, the power difference PD_g is expressed as (4.7), so that the reserve capacity is a positive number and thus evaluation indices can be represented as a universal form.

$$PD_g(\mathbf{u}_g, \tau_g, t) = \begin{cases} P_g^0(\mathbf{u}_g, \tau_g, t) - P_g(\mathbf{u}_g, \tau_g, t)\ ,\ load - reduction \\ P_g(\mathbf{u}_g, \tau_g, t) - P_g^0(\mathbf{u}_g, \tau_g, t)\ ,\ load - increase \end{cases} \quad (4.7)$$

where $P_g^0(\mathbf{u}_g, \tau_g, t)$ and $P_g(\mathbf{u}_g, \tau_g, t)$ are the aggregated power of group g before and after the changes of the set point temperature, respectively. Equation (4.7) is not expressed in the form of absolute value because the rebound load may exceed the aggregate power at the initial state. Hence, evaluation indices represented by the PD_g can be applied to both the load-reduction operation during the reserve deployment period and the load-increase operation during the recovery period, respectively. A general evaluation framework for the control of the lead-lag rebound is developed based on the dynamics of power difference after the changes of set point temperature, which is shown by Fig. 4.3.

4.3.2 Evaluation of the Operating Reserve Provided by ACs on the Capacity Dimension

(1) **Reserve Capacity (RC)**

The aggregate power fluctuates in nature due to the cyclic operation characteristics of ACs. Let $PD_g^{\max}$ denotes the maximum power difference of group g during reserve deployment period, the valid reserve deployment is then defined as the threshold

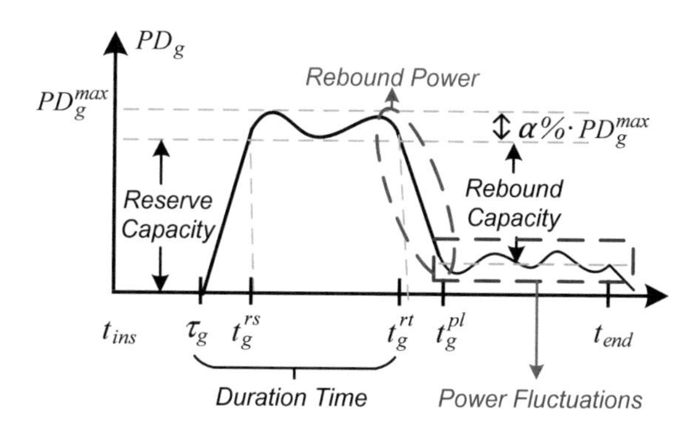

Fig. 4.3 Power difference of the aggregate power before and after the changes of the set point temperature

between $(1 - \alpha\%) \cdot PD_g^{\max}$ and $PD_g^{\max}$. t_g^{rt} and t_g^{rs} are the time instant corresponding to two endpoints of the defined threshold. Hence, the aggregate reserve capacity (RC) is:

$$RC_g(\mathbf{u}_g, \tau_g) = PD_g^{\max}(\mathbf{u}_g, \tau_g) - PD_g^{\max}(\mathbf{u}_g, \tau_g) \times \alpha\% \tag{4.8}$$

(2) **Rebound Capacity (BC)**

As is illustrated in Fig. 4.2, the power difference declines because of the rebound. The decline process stops when the aggregate power of the ACs enter the steady state. Denote t_g^{pl} as the end of the rebound process, which is defined as the time instant when power difference stops declining. The rebound capacity (BC) is the difference between the reserve capacity and power difference at t_g^{pl}:

$$BC_g(\mathbf{u}_g, \tau_g) = RC_g(\mathbf{u}_g, \tau_g) - PD_g(\mathbf{u}_g, \tau_g, t_g^{pl}) \tag{4.9}$$

(3) **Power Volatility (PV)**

Standard deviation (SD) is adopted to represent the fluctuations of aggregate power. The standard deviation of the variable $x(t)$ during the time period $[t_1, t_2]$ is defined as:

$$\underset{t_1 \to t_2}{SD}(x(t)) = \sqrt{\left[\int_{t_1}^{t_2} (x(t) - \bar{x}(t))^2 \Delta t\right] / (t_2 - t_1)} \tag{4.10}$$

where $\bar{x}(t)$ is the mean value of the variable $x(t)$ from t_1 to t_2.

The standard deviation of the power difference after the power spike caused by the rebound peak represents the power fluctuations caused by the AC load control. Power volatility (PV) is defined as the ratio of the standard deviation to the initial aggregate power at the time instant of reserve deployment:

$$PV_g = \underset{t_g^{pl} \to t_{end}}{SD} (PD_g(\mathbf{u}_g, \tau_g, t)) / P_g(\mathbf{u}_g, \tau_g, \tau_g) \tag{4.11}$$

4.3.3 Evaluation of the Operating Reserve Provided by ACs on the Time Dimension

(1) **Duration Time (DT)**

Duration time (DT) is the period that reserve service providers maintain the required reserve capacity RC^*. Traditionally, DT is calculated as the period between the reserve deployment and the recovery [32]. However, ACs cannot maintain RC^* once the lead rebound occurs. In this case, DT is the period within the defined reserve capacity threshold in (4.8). Therefore, DT is quantified according to the rebound time instant:

$$DT_g = \begin{cases} t_g^{rt} - \tau_g \ , \ t_g^{rt} < \tau_g + DT^* \\ t_{end} - \tau_g \ , \quad otherwise \end{cases} \tag{4.12}$$

(2) **Ramp Rate (RR)**

Ramp time (RT) is the period that the reserve service providers control their output to the required reserve capacity:

$$RT_g = t_g^{rs} - \tau_g \tag{4.13}$$

Ramp rate (RR) is the speed that the reserve service providers control their output to the required value [32]. RR is defined as the ratio of the reserve capacity to the ramp time:

$$RR_g(\mathbf{u}_g, \tau_g) = RC_g(\mathbf{u}_g, \tau_g)/RT_g \tag{4.14}$$

4.4 Sequential Dispatch Strategy of ACs for Providing Operating Reserve with Multiple Duration Time

4.4.1 The Interactions Among the System Operator, Aggregators and Consumers

Three types of entities involve in the provision of operating reserve by ACs, i.e., the system operator, the aggregators and the consumers [39]. The interactions among the entities are shown in Fig. 4.4.

The role of system operator (e.g., the Independent System Operator (ISO) in the United States, Transmission System Operator (TSO) in the European Commission or the grid company in China) is to operate the transmission system [39, 40]. The system operator will run multiple DR programs to motivate consumers for benefiting

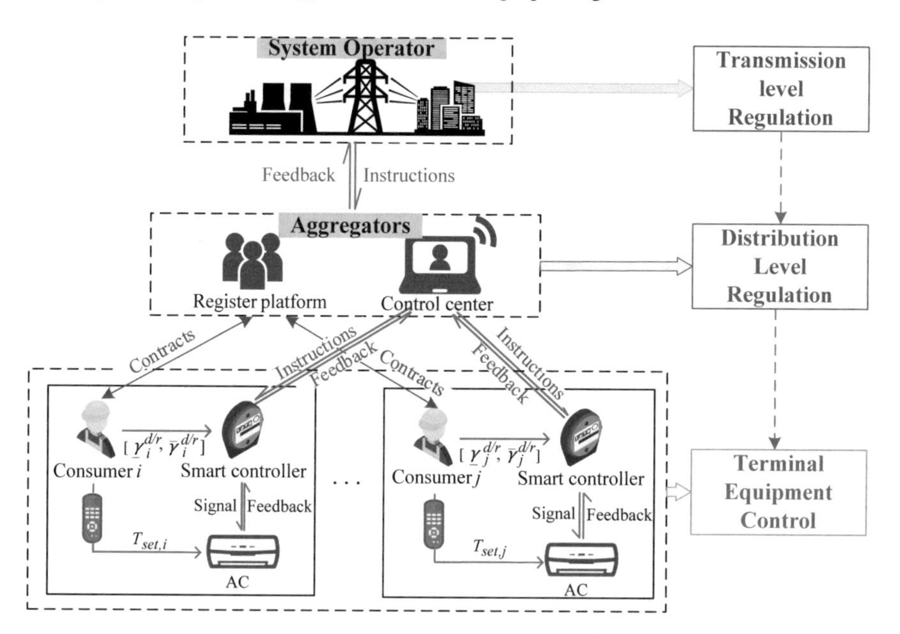

Fig. 4.4 The interactions among the system operator, aggregators and consumers

transmission system operations by actively reducing or increasing electricity consumption [41]. Typically, ancillary service market programs allow consumers to act as reserve service providers for providing operating reserve on equal terms with the generators [41]. During the operating hours, in order to correct the imbalance between generation and demand, the system operator will instruct reserve service providers about when and how much the operating reserves to be deployed and recovered [35].

Small consumers, including the owners of ACs, are grouped by aggregators to bid in the market because the limited capacity of individual consumer cannot fulfill the requirement on the minimum amount of bids in the spot market [39]. Depending on the specific design and structure of a DR program, the aggregators can be distribution system operators, load-serving entities, or DR providers [41]. By signing contracts with consumers, aggregators can control ACs at the permitted periods [35]. If the bids from aggregators for the provision of operating reserve are accepted in the market, the aggregators will respond to the instructions from the system operator by regulating the controllable ACs with the proposed sequential dispatch strategy during the operating hours. On this basis, aggregators will send instruction signals to the smart controllers of the ACs to be dispatched [38].

It is assumed that the ACs are installed with smart controllers, which share the functions of communication, sensor and control [42]. The smart controllers enable consumers to easily set the parameters, such as the temperature ranges, controllable periods and control modes [9, 43]. After receiving the instructions from aggregators, the smart controllers will then control ACs with local embedded control strategy according to the parameters set by consumers.

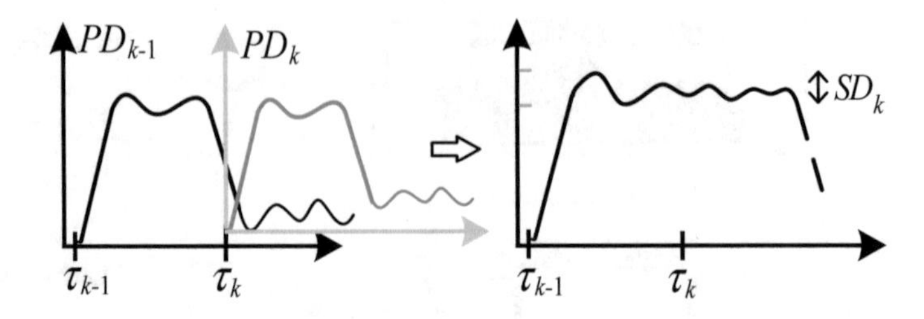

Fig. 4.5 Principle of the sequential dispatch strategy

4.4.2 Sequential Dispatch Strategy of ACs to Mitigate the Lead-Lag Rebound Effect

In order to mitigate the rebound load and thus enable the flexible control of duration time, a sequential dispatch strategy of ACs is proposed. The principle of sequential dispatch strategy is shown in Fig. 4.5. The rebound capacity can be an updated required reserve capacity. If ACs are divided into several groups and if the reserve capacity of each group is equal to the rebound capacity caused by the previous group, then the rebound load of the previous group is mitigated entirely.

The first group of ACs is dispatched when receiving reserve deployment instruction at t_{ins}. The required reserve capacity of the first group is equal to the value instructed by the system operator:

$$RC_1^*\left(\boldsymbol{\theta}^+, \boldsymbol{\theta}^-, \boldsymbol{\gamma}_1^{d/r}, \mathbf{S}_1, t_{ins}\right) = RC^* \tag{4.15}$$

The latter groups of ACs are dispatched continuously to mitigate the rebound load caused by the previous group.

$$RC_k^*\left(\boldsymbol{\theta}^+, \boldsymbol{\theta}^-, \boldsymbol{\gamma}_k^{d/r}, \mathbf{S}_k, \tau_k\right) = BC_{k-1}\left(\boldsymbol{\theta}^+, \boldsymbol{\theta}^-, \boldsymbol{\gamma}_k^{d/r}, \mathbf{S}_{k-1}, \tau_{k-1}\right) \tag{4.16}$$

The process of sequential dispatch terminates when the rebound capacity is sufficiently small. Note that the stabilized aggregate power also fluctuates within a range and the fluctuations will increase with larger number of ACs. Therefore, the termination condition of the sequential-dispatching process is set as a value proportional to the aggregate power at t_{ins}. β denotes the proportional coefficient of the termination condition, and its value can be set according to the power volatility PV in steady state. Suppose the rebound load has been mitigated entirely when:

$$BC_k \le \beta\% \cdot P_k(\mathbf{u}_k, \tau_k, t_{ins}) \tag{4.17}$$

As illustrated in Figs. 4.1 and 4.2, the rebound load and the fluctuations of aggregate power are highly relevant to the temporary synchronization of the ACs. Existing methods on avoiding the synchronization of ACs can be classified into two categories.

The first is to track power profiles by subtle changes of the set point temperature in real time based on two-way communication between control center and ACs [34, 44, 45]. In [34, 44], synchronization of homogenous ACs is governed by broadcasting subtle temperature set point changes as the output signal of a feedback-controller. Reference [45] includes AC parameter heterogeneity by controlling the on/off states of ACs based on the state bin transition model. However, this method requires careful tuning of parameters for specific scenarios and is difficult to be applied to the control of ACs with changes of set point temperature. Therefore, this approach is more suitable for the provision of load following or regulation services, which usually accommodate ACs with fast communication equipment and control ACs through subtle changes of set point temperature frequently (e.g., minute-to-minute changes of set point temperature smaller than 0.1 °C [34]).

The second is to implement the shift in the set point temperature according to safe protocols embedded in the smart controller of each individual AC [46–49]. References [46, 47] propose several safe protocols to generate different power pulse shapes and avoid the sudden changes of ACs' state, which have been proved to effectively avoid the synchronization for both heterogeneous and homogeneous ACs with different changes of set point temperature [48, 49]. Therefore, such approach based on safe protocols is more suitable for providing operating reserve in this chapter, which controls heterogeneous ACs with larger changes of set point temperature for only two times (one at the reserve deployment time instant and the other at the recovery time instant). Among all the safe protocols proposed in [46, 47], the safe protocol-2 (SP-2) can reduce power fluctuations to the lowest level and is adopted in this chapter.

Implementation of the sequential dispatch strategy coordinated with the SP-2 is illustrated in Fig. 4.6. The smart controller of the i-th AC is abbreviated to SC_i. Firstly, according to the requirements of operating reserve on the capacity dimension and the time dimension, the sequential-dispatching controller optimizes the use of all the controllable ACs under an aggregator with the proposed sequential dispatch strategy. The sequential-dispatching controller will then send signals to the smart controllers of ACs in group k about the reserve deployment time instant and the changes of set point temperature. Secondly, after receiving the instructions from the sequential-dispatching controller, the control of ACs will follow SP-2, which is embedded in the smart controller of an individual AC. In this way, the fluctuations of aggregate power can be largely reduced without adding additional computational burden between aggregators and consumers. Thirdly, the detailed information of the reserve deployment is feedback to aggregators.

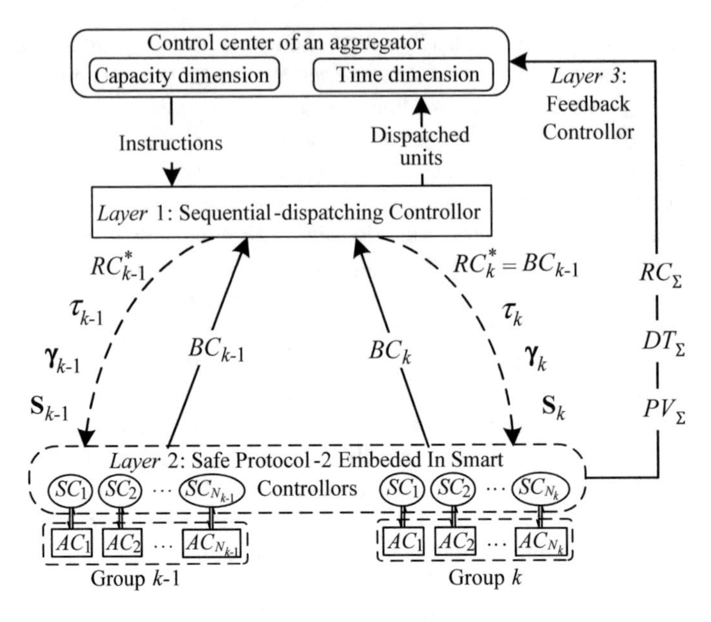

Fig. 4.6 Three-layer structure for the implementation of sequential dispatch strategy

4.4.3 Capacity-Time Co-optimization of Sequential Dispatch Process During the Reserve Deployment Period

The dispatch time instant and available reserve capacity of each AC group are co-optimized to make full use of ACs' potential for the provision of operating reserve. For clarity, the subscript d and r denote the parameters of reserve deployment period and the recovery period, respectively. Figure 4.5 shows that the dispatch time instant τ_k^d of group k greatly influences the performance of dispatching group k to mitigate the rebound load caused by group k-1. For example, if group k is dispatched too early, then the rebound load of group k and group k-1 will accumulate, resulting in higher rebound load. By contrast, if group k is dispatched too late, then the aggregate power will still rebound until group k is dispatched. Therefore, the deployment time instant τ_k^d of group k is optimized to minimize the deviation between the actual power difference and the required value RC^*, that is,

$$\min_{\tau_k^d} \underset{t_{ins}\rightarrow t_k^{rt}}{SD} \left\{ \left(\sum_{j=1}^{k} PD_j(\boldsymbol{\theta}^+, \boldsymbol{\theta}^-, \boldsymbol{\gamma}^d, \mathbf{S}_j^d, \tau_j^d, t) \right) - RC^* \right\} \tag{4.18}$$

The largest available reserve capacity of an AC group can be calculated when the set point temperature of all the ACs are changed to the bound set by consumers. Aggregators tend to make the most of the available reserve capacity to earn more benefits. Therefore, the selection of ACs $\mathbf{S}_k^d$ in group k and the corresponding changes of set point temperature $\boldsymbol{\gamma}_k^d$ are optimized to follow the requirements of reserve capacity with minimum available reserve capacity:

$$\min_{\mathbf{S}_k^d, \boldsymbol{\gamma}_k^d} RC_k^d(\boldsymbol{\theta}^+, \boldsymbol{\theta}^-, \widehat{\gamma}^d, \mathbf{S}_k^d, \tau_k^d) \tag{4.19}$$

where $\widehat{\gamma}^d$ is the maximum changes of the set point temperature within the bound set by consumers. In other words, $\widehat{\gamma}^d = \bar{\boldsymbol{\gamma}}^d$ for load-reduction in summer or load-increase in winter, $\widehat{\gamma}^d = \underline{\boldsymbol{\gamma}}^d$ for load-increase in summer or load-reduction in winter.

Equations (4.18) and (4.19) form a bi-level optimization:

$$\min_{\tau_k^d} \underset{t_{ins} \to t_k^{rt}}{SD} \left\{ \left(\sum_{j=1}^{k} PD_j(\boldsymbol{\theta}^+, \boldsymbol{\theta}^-, \boldsymbol{\gamma}_j^d, \mathbf{S}_j^d, \tau_j^d, t) \right) - RC^* \right\} \tag{4.20}$$

s.t.

$$\tau_k^d > \tau_{k-1}^d \tag{4.21}$$

$$\tau_k^d < t_{ins} + DT^* \tag{4.22}$$

$$[\mathbf{S}_k^d, \boldsymbol{\gamma}_k^d] = \arg\min RC_k^d(\boldsymbol{\theta}^+, \boldsymbol{\theta}^-, \widehat{\gamma}^d, \mathbf{S}_k^d, \tau_k^d) \tag{4.23}$$

s.t.

$$RC_k^{d,*}\left(\boldsymbol{\theta}^+, \boldsymbol{\theta}^-, \boldsymbol{\gamma}_k^d, \mathbf{S}_k^d, \tau_k^d\right) = BC_{k-1}^d\left(\boldsymbol{\theta}^+, \boldsymbol{\theta}^-, \boldsymbol{\gamma}_{k-1}^d, \mathbf{S}_{k-1}^d, \tau_{k-1}^d\right) \tag{4.24}$$

$$\underline{\boldsymbol{\gamma}}^d \le \boldsymbol{\gamma}_k^d \le \bar{\boldsymbol{\gamma}}^d \tag{4.25}$$

$$\sum_{j=1}^{k} \sum_{i=1}^{N^{\max}} s_{j,i}^d \times v_i \le \sum_{i=1}^{N^{\max}} v_i \tag{4.26}$$

The high-level problem (4.20)–(4.22) optimize the deployment time instant of group k to minimize the deviation between the actual power difference and the required value. Equation (4.21) ensures that the deployment time instant of the current group is later than the previous group, while (4.22) limits the dispatch operation within the required duration time DT^*. The low-level problem (4.23)–(4.26) optimize the selection of ACs in group k and the corresponding changes of set point temperature to follow the requirements of reserve capacity with minimum available reserve capacity of ACs. Equation (4.24) constraints the reserve capacity of group k according to the rebound capacity of the previous group. Equation (4.25) limits the changes of the set point temperature within the range set by the consumers. Equation (4.26) constraints the total number of dispatched ACs within maximum number of available

ACs. The bi-level optimization formed by (4.20)–(4.26) is a mixed integer nonlinear bi-level programming problem. Genetic algorithm (GA) provides a flexible modeling framework that allows considering the nonlinearities and non-convexities associated with the mixed integer nonlinear bi-level programming problem [50]. Therefore, GA is applied to solve the bi-level optimization formed by (4.20)–(4.26) in this chapter.

The optimization of the low-level problem (4.23)–(4.26) cannot continue when (4.24) and (4.26) cannot be satisfied at the same time. In other words, the remaining ACs are not adequate to mitigate the rebound load caused by the previous group entirely. K denotes the number of all the dispatched AC groups during the reserve deployment period. In this case, the units belonging to the K-th group are selected as the remaining ACs:

$$\mathbf{S}_K^d = \mathbf{V} - \sum_{j=1}^{K-1} \mathbf{S}_j^d \tag{4.27}$$

After all the available ACs have been dispatched, the sequential dispatch process is terminated. Hence, the duration time is determined by the rebound time instant of group k and is represented by:

$$DT = t_K^{rt} - t_{ins} \tag{4.28}$$

4.4.4 *Capacity-Time Co-optimization of Sequential Dispatch Process During the Recovery Period*

ACs will be recovered to the initial states when the duration time reaches the required value. Sequential dispatch strategy can also be utilized to mitigate the lag rebound. Since the ACs to be recovered are the same as those dispatched during the reserve deployment period, there is no need to conduct the optimization revealed in (4.23)–(4.26). Equation (4.25) has ensured that the changes of set point temperature are within the range set by consumers and therefore the consumers' basic comfort levels can be guaranteed. However, consumers' thermal comfort levels will still decrease with longer deployment duration or larger changes of the set point temperature [42]. In order to avoid the further dissatisfaction of consumers, ACs with the lowest comfort levels should be recovered earlier. $DT_i^{\max}$ denotes the maximum allowable control duration of the i-th AC according to the contract with the aggregator. The objective function to determine ACs in the q-th group is then represented by (4.29), so that the ACs with the lowest thermal comfort levels are selected to be recovered.

$$\min_{\mathbf{S}_q^r} \sum_{i=1}^{N^{\max}} \sum_{m=1}^{K} \left(1 - \frac{\tau_q^r - \tau_m^d}{DT_i^{\max}} \cdot \frac{\gamma_{m,i}^d}{\widehat{\gamma}_i^d} \right) \cdot s_{m,i}^d \cdot s_{q,i}^r \tag{4.29}$$

Similar with the reserve deployment process, the optimization of the recovery time instant τ_q^r and the selection of ACs S_q^r of group q form bi-level optimization:

$$\min_{\tau_q^r} \underset{t_{end} \to t_q^{rt}}{SD} \left\{ \left(\sum_{j=1}^{q} PD_j(\boldsymbol{\theta}^+, \boldsymbol{\theta}^-, \boldsymbol{\gamma}_j^r, \mathbf{S}_j^r, \tau_j^r, t) \right) - RC^* \right\} \tag{4.30}$$

s.t.

$$\tau_q^r > \tau_{q-1}^r \tag{4.31}$$

$$\tau_q^r < t_{end} + \mu \tag{4.32}$$

$$\mathbf{S}_q^r = \arg\min \sum_{i=1}^{N^{\max}} \sum_{m=1}^{K} \left(1 - \frac{\tau_q^r - \tau_m^d}{DT_i^{\max}} \cdot \frac{\gamma_{m,i}^d}{\widehat{\gamma}_i^d} \right) \cdot s_{m,i}^d \cdot s_{q,i}^r \tag{4.33}$$

s.t.

$$\boldsymbol{\gamma}_q^r = -\boldsymbol{\gamma}_K^d \tag{4.34}$$

$$\left| RC_q^{r,*}(\boldsymbol{\theta}^+, \boldsymbol{\theta}^-, \boldsymbol{\gamma}_q^r, \mathbf{S}_q^r, \tau_q^r) - BC_{q-1}^r(\boldsymbol{\theta}^+, \boldsymbol{\theta}^-, \boldsymbol{\gamma}_{q-1}^r, \mathbf{S}_{q-1}^r, \tau_{q-1}^r) \right| < \underline{p} \tag{4.35}$$

$$\underline{p} = \min\left(p_i | \sum_{m=1}^{K} s_{m,i}^d \neq \sum_{l=1}^{q-1} s_{l,i}^r, i = 1, 2, \cdots, N^{\max} \right) \tag{4.36}$$

$$\mathbf{S}_q^r \leq \sum_{m=1}^{K} \mathbf{S}_m^d - \sum_{l=1}^{q-1} \mathbf{S}_l^r \tag{4.37}$$

The high-level problem (4.30)–(4.32) optimize the recovery time instant of group q, which is similar to (4.20)–(4.22). μ denotes the duration in which the recovery process has to be finished. Therefore, the high-level optimization is limited within the period $[t_{end}, t_{end} + \mu]$. The low-level problem (4.33)–(4.37) optimize the selection of ACs in group q so that the ACs with the lowest thermal comfort levels are recovered earlier. Equation (4.34) resets the set point temperature to its original value. In other words, the aggregate power of ACs cannot be flexibly controlled with changing set point temperature. Therefore, (4.35) represents that the rebound load of group q-1 is mitigated entirely by the reserve capacity of group q when their difference is smaller than $\underline{p}$, which denotes the minimum power of the remaining ACs and is calculated as (4.36). Equation (4.37) constraints ACs in group q within dispatched ACs that have not been recovered.

Considering that the operation process of an individual AC is described by the general state model for TCLs in (4.1), the proposed method may be applied to other TCLs, such as refrigerators, heat pump space heaters and electric water heaters [51].

Statistical data have shown that TCLs account for 48, 35, 40 and 51% of residential electricity consumption in the U.S. [52], the UK [53], Australia [54] and China [55], respectively. Therefore, the proposed method could be applied to different regions considering the widespread of TCLs. Among all the common TCLs, the cycle time of AC is relatively short, leading to a short deployment duration constrained by the lead rebound. Therefore, AC is taken as a typical type of TCL to show the effect of the lead-lag rebound and the effectiveness of the proposed strategy.

4.5 Case Studies and Simulation Results

Case studies are conducted to validate the effectiveness of the sequential dispatch strategy for providing operating reserve with various duration time. First, the potential of ACs for the provision of operating reserve is evaluated by the sequential dispatch of ACs without the recovery program. Second, the performance of the sequential dispatch and recovery strategy is verified by the reserve deployment of ACs with various required reserve capacity and duration time. Furthermore, operating reserve provided by ACs are used to relieve congestion resulted from system peak load in IEEE-30-bus test system. On this basis, different dispatch strategies of ACs in existing research studies are compared to verify the necessity of mitigating lead-lag rebound with the proposed sequential dispatch strategy.

An aggregator with controllable ACs in a residential area in summer, which is when all of the ACs operate in cooling mode, is modeled. The simulation parameters are illustrated in Table 4.1. The operation parameters of ACs are generated from a pilot study which obtained spinning reserve from responsive air conditioning loads at a motel over a year [56]. The coefficient of performance COP_i of the i-th AC is set according to [57], which generates this parameter from AC operating data published by Bosch Termoteknik. Thermal parameters of rooms are set according to [34], which lists the bulk thermal properties of buildings from measurement data in experimental studies. It is assumed that aggregators are permitted to control the ACs for at least one hour by contract. The ambient temperature is set as 32 °C. The temperature dead band and the maximum changes of set point temperature are set to 1 °C and 2 °C, respectively. The proportional coefficient $\beta\%$ in (4.17) is set as 10%.

4.5.1 *Evaluation of ACs' Potential for the Provision of Operating Reserve*

This case simulates the reserve deployment of ACs without the recovery program, through which the maximum duration time corresponding to the required reserve capacity RC^* can be observed. ACs are dispatched to provide load-reduction service at the time 16:00. The dynamics of the sequential dispatch process of N^{max} ACs are

Table 4.1 AC physical parameters

Parameters	Descriptions	Values	Units
A_i	Room area	$N(20, 25)$ [56]	m^2
C_i	Thermal capacity	$0.015 \cdot A_i$ [34]	kWh/°C
R_i	Thermal resistance	$100 \cdot A_i^{-1}$ [34]	°C/kW
COP_i	Coefficient of performance	$-0.0384\|\theta_a(t) - \theta_i(t)\|$ +3.9051[56]	/
p_i	Input power	$U(40 \cdot A_i, 70 \cdot A_i)$ [56]	W
T_{set}	Set point temperature	$U(23, 28)$	°C

Normal distribution with the mean value of μ and the standard deviation of σ is abbreviated to $N(\mu, \sigma^2)$; uniform distribution with the minimum and maximum value of a and b, respectively, is abbreviated to $U(a,b)$

demonstrated in Fig. 4.7. The aggregate power of all dispatched ACs is labeled as AG-Total, below which the aggregate power of the g-th group is labeled as AG-g. The number of ACs and the dispatch time instant of the g-th group τ_g are shown in Table 4.2.

The curve of AG-Total in Fig. 4.7a shows that aggregate power of all the dispatched ACs decreases from 12 to 7 MW after receiving the reserve deployment signal and maintains at 7 MW since then. Hence, the lead rebound is eliminated entirely after the sequential dispatch of five groups of ACs (shown by AG-1 to AG-5) when N^{max} is 60,000. By contrast, RC^* in Fig. 4.7a is the same with that in Fig. 4.7b, while N^{max} of the latter is 35,000 smaller than the former. Table 4.2 shows that the dispatch results of AG-1 and AG-2 are the same. However, all the remaining ACs in Fig. 4.7b are utilized in AG-3 and the curve of AG-Total in Fig. 4.7b shows that the aggregate power of dispatched ACs rebounds at around 16:45. Hence, 25,000 controllable ACs are not sufficient to mitigate the rebound load entirely and the maximum duration time is only about 45 min. On the other hand, N^{max} in Fig. 4.7a and c is the same and equals to 60,000, while RC^* of the latter is 10 MW higher than the former. The

Table 4.2 Dispatch results of ACs during the reserve deployment period

RC^*/N^{max}	AG-No.	Number	τ_g	AG-No.	Number	τ_g
5 MW/60,000	1	14,143	16:00	2	8,143	16:22
	3	5,663	16:41	4	3,518	17:03
	5	2,524	17:24	–	–	–
5 MW/25,000	1	14,143	16:00	2	8,143	16:22
	3	2,714	16:36	–	–	–
15 MW/60,000	1	42,279	16:00	2	17,721	16:19

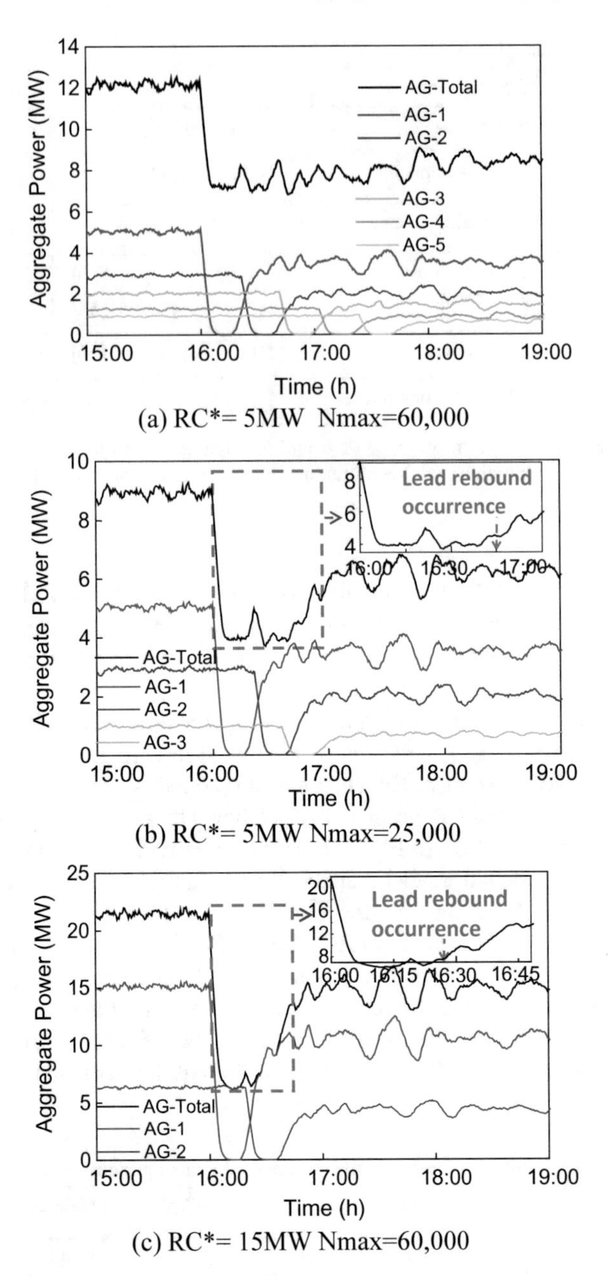

Fig. 4.7 Operating reserve provided by ACs when receiving reserve deployment signal at 16:00

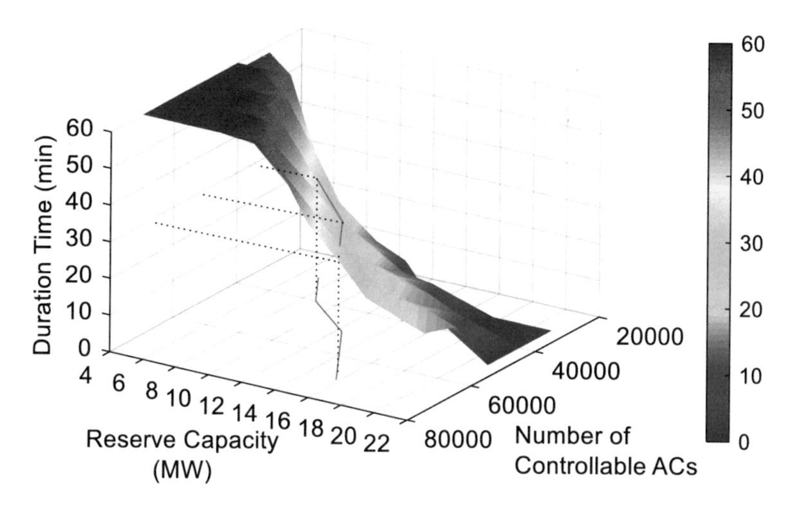

Fig. 4.8 Maximum duration time corresponding to different numbers of controllable ACs and different required reserve capacity

curve of AG-Total in Fig. 4.7c shows that the aggregate power of dispatched ACs rebounds at around 16:25. Figure 4.7c and Table 4.2 show that after the dispatch of two groups (AG-1 and AG-2), all the controllable ACs are utilized. Therefore, 60,000 controllable ACs are not sufficient to mitigate the rebound load entirely when RC^* is 15 MW and the maximum duration time is only about 25 min. Consequently, there exists constraints between the feasible reserve capacity and feasible deployment duration, both of which are also limited by total number of controllable ACs.

In order to evaluate ACs' potential for the provision of operating reserve, the maximum duration time DT^{max} corresponding to different numbers of controllable ACs and RC^* is simulated, as demonstrated in Fig. 4.8. In this way, the feasibility ranges of operating reserve are the areas below the surface in Fig. 4.8. The simulation cases are conducted on a PC with Intel 2.3 GHz 2-core processor (4 MB L3 cache), 8 GB memory. The computational time of the sequential dispatch and recovery process when N^{max} controllable ACs are required to provide reserve capacity of RC^* for DT^{max} is shown in Fig. 4.9. Since all the ACs are assumed to be controllable for at least one hour, the longest deployment duration is set as 60 min, after which the sequential dispatch procedure will stop. For a given number of ACs, if the maximum duration time in Fig. 4.8 equals to 60 min, the computational time in Fig. 4.9 increases with RC^* because more ACs are dispatched with larger RC^*. On the other hand, if the maximum duration time in Fig. 4.8 is smaller than 60 min, all the controllable ACs are dispatched to fulfill the requirement of RC^*. The computational time in Fig. 4.9 decreases with RC^* because less groups of ACs will be dispatched/recovered with larger RC^*. Hence, the computational time reaches the maximum value when DT^* has just reached 60 min. The maximum computational time in Fig. 4.9 is 248 s, which is corresponding to the provision of 12 MW reserve capacity for 60 min with 70,000 ACs.

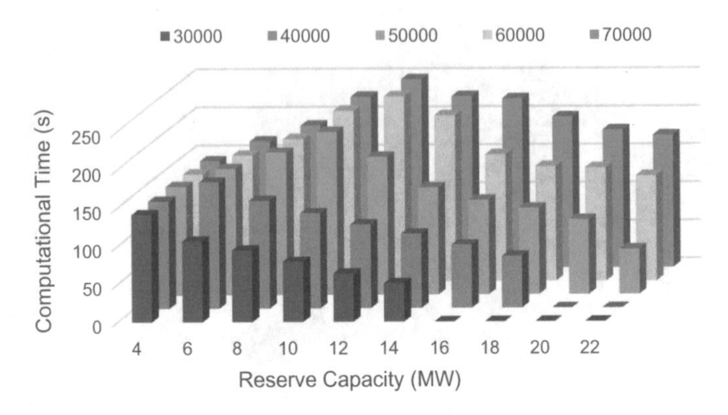

Fig. 4.9 Computational time corresponding to different numbers of controllable ACs and different required reserve capacity

For a given number of ACs, the maximum duration time decreases with the increase of RC^*. Consequently, reserve capacity in Fig. 4.8 reaches the maximum value when the maximum duration time equals to the required value DT^*. Assume that the required duration time is 30 min [31], the maximum reserve capacity corresponding to the number of controllable ACs is plotted on the bottom of Fig. 4.8, which shows that the maximum reserve capacity is 14.09 MW when there are 60,000 ACs. Hence, when RC^* is lower than 14.09 MW, ACs can fulfill the requirements of duration time, as illustrated in Fig. 4.7a. By contrast, when RC^* is 15 MW, the duration time is not enough, as illustrated in Fig. 4.7c. Therefore, the maximum reserve capacity can be effectively evaluated according to the expected duration time and total number of controllable ACs. Such evaluation is helpful for the selection of RC^* and DT^* in the following cases.

4.5.2 Provision of Operating Reserve with Various Duration Time and Reserve Capacity

In this case, the ACs are required to provide operating reserve with a specified reserve capacity RC^* and a specified duration time DT^*. Hence, ACs should ensure that the lead rebound is mitigated entirely during DT^*, after which the recovery program will be triggered. The maximum number of controllable ACs is set as 60,000. The dynamics of reserve deployment with the recovery process are shown in Fig. 4.10, in which the dispatch results during the reserve deployment period and the recovery period are separated by the dotted line. The room temperature profile corresponding to the load control in Fig. 4.10a–e are shown in Fig. 4.11a–e, respectively. The indices for the simulated case are presented in Table 4.3. Standard deviation *SD* and power volatility *PV* are the values for the load recovery process. The threshold of valid reserve capacity calculated by (4.8) is set between $0.9 \cdot PD_g^{\max}$ and $PD_g^{\max}$. The

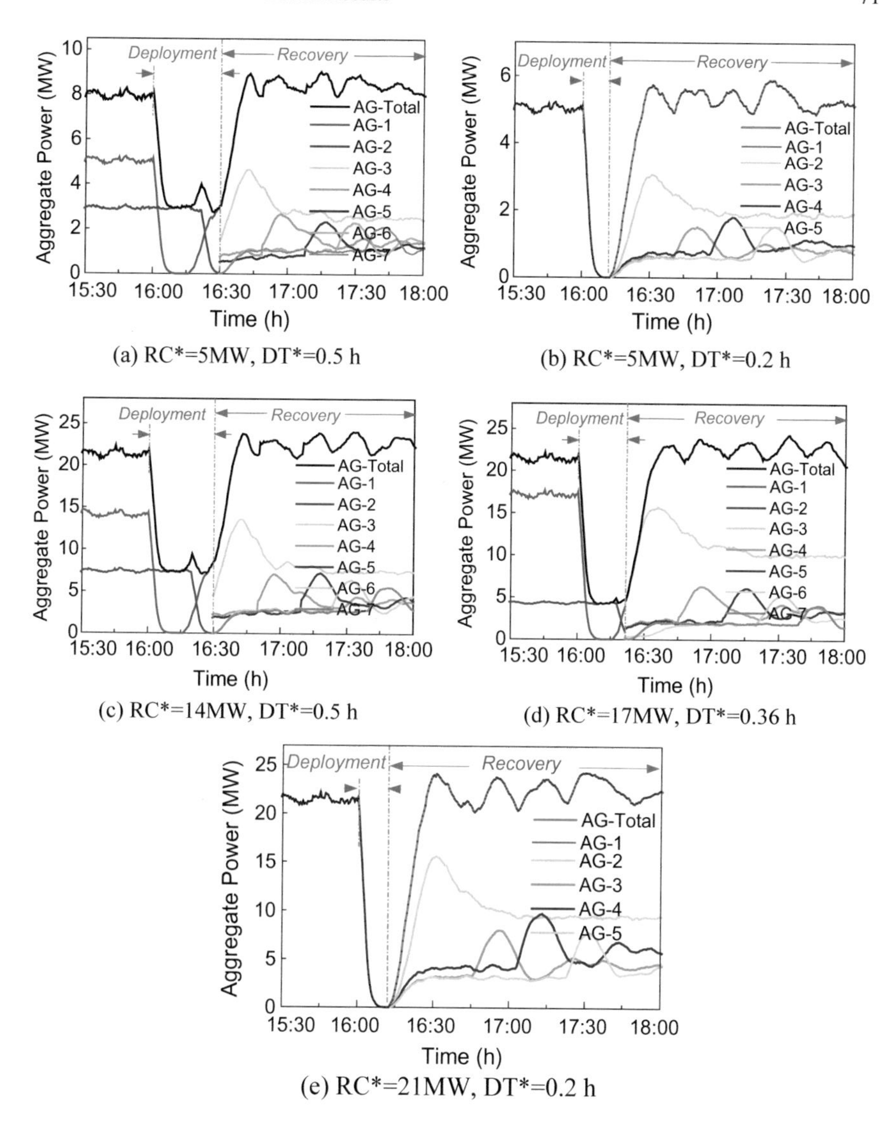

Fig. 4.10 Sequential dispatch and recovery of ACs for the provision of operating reserve with various RC^* and DT^*

number of ACs and the dispatch/recovery time instant of the g-th group τ_g are shown in Table 4.4.

Aggregate power of all dispatched ACs maintains at the reduced value during the reserve deployment period and returns the initial value steadily after receiving the recovery signal, as is illustrated by the curve of AG-Total in Fig. 4.10. This means that both the lead rebound and lag rebound are mitigated entirely by dispatching different AC groups in sequence. Dynamics of reserve deployment are various with different

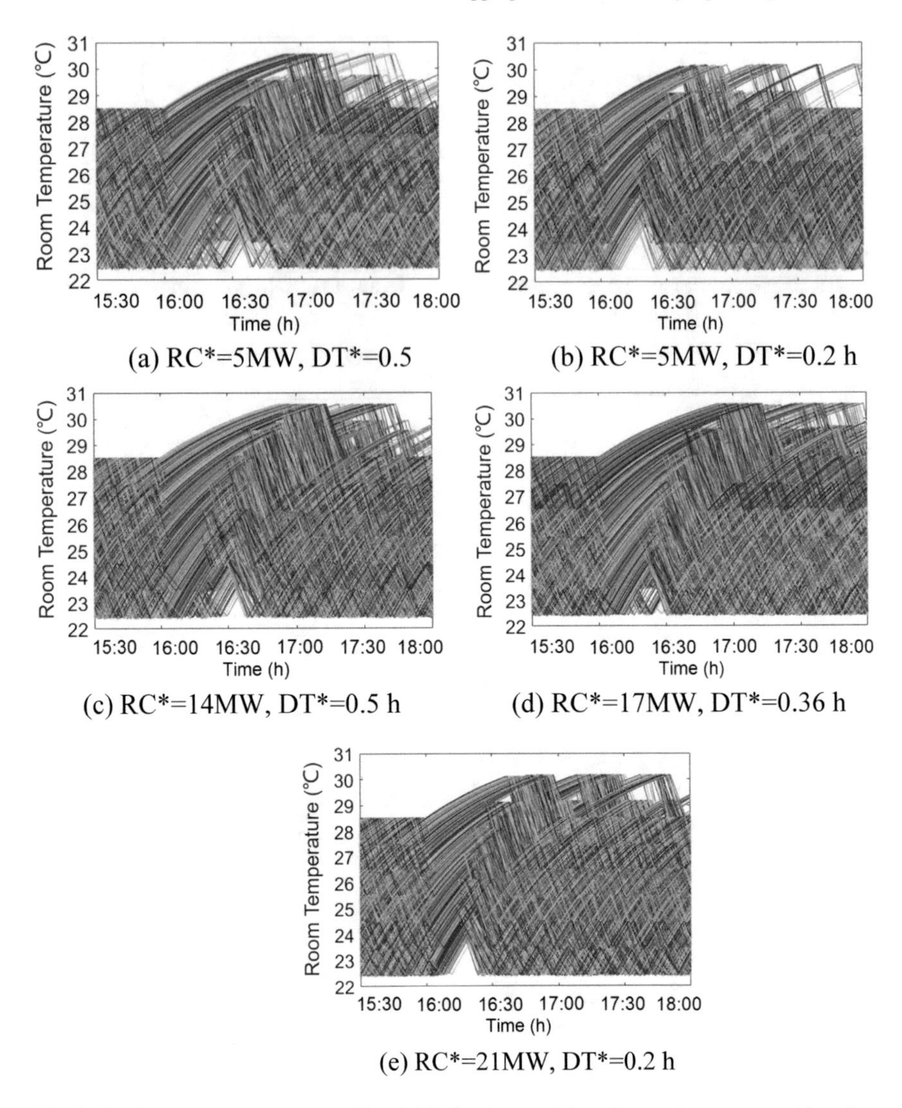

Fig. 4.11 The room temperature profile of ACs for the provision of operating reserve with various RC^* and DT^*

DT^* and RC^*. On the one hand, RC^* in Fig. 4.10a and b is the same (5 MW), while DT^* of the latter is 0.3 h lower than the former. Table 4.4 shows that only 14,143 ACs dispatched at 16:00 are enough to realize the deployment duration of 0.2 h in Fig. 4.10b, while another 8,143 ACs are dispatched at 16:21 in Fig. 4.10a to extend the deployment duration to 0.5 h. Hence, ACs are divided into more groups to be recovered in Fig. 4.10a, leading to the decrease of PV by 2.11% (= 5.86 − 3.75%) compared to that in Fig. 4.10b, as shown in Table 4.3. Figure 4.11a–b show that

Table 4.3 Indices of the operating reserve with various RC* and DT*

Instruction (RC^*/DT^*)	SD (MW)	PV (%)	RT_d (min)	RT_r (min)	RR_d (MW/min)	RR_r (MW/min)
5 MW/0.5 h	0.30	3.75	5.03	8.99	0.99	0.56
5 MW/0.2 h	0.29	5.86	4.92	12.08	1.02	0.41
14 MW/0.5 h	0.83	3.96	5.98	8.52	2.81	1.64
17 MW/0.36	0.88	4.19	5.40	9.67	3.15	1.76
21 MW/0.2 h	1.20	5.71	5.11	12.19	4.11	1.74

Table 4.4 Dispatch results of ACs during the reserve deployment period and the recovery period

RC^*/DT^*	Period	AG-No.	Number	τ_g	AG-No.	Number	τ_g
5 MW/0.5 h	Deployment	1	14,143	16:00	2	8,143	16:21
	Recovery	3	7,329	16:30	4	4,031	16:49
		5	3,435	17:08	6	3,804	17:21
		7	3,687	17:34	–	–	–
5 MW/0.2 h	Deployment	1	14,143	16:00	–	–	–
	Recovery	2	6,013	16:12	3	2,605	16:40
		4	3,050	17:12	5	2,475	17:14
14 MW/0.5 h	Deployment	1	39,431	16:00	2	20,569	16:22
	Recovery	3	20,673	16:30	4	10,099	16:50
		5	10,104	17:09	6	9,868	17:22
		7	9,256	17:39	–	–	–
17 MW/0.36 h	Deployment	1	48,014	16:00	2	11,986	16:18
	Recovery	3	28,166	16:22	4	9,299	16:48
		5	8,601	17:08	6	7,446	17:23
		7	6,498	17:42	–	–	–
21 MW/0.2 h	Deployment	1	60,000	16:00	–	–	–
	Recovery	2	24,354	16:11	3	10,942	16:45
		4	14,092	17:02	5	10,612	17:22

the reduced deployment duration in Fig. 4.10b leads to the decreased changes of set point temperature by approximately 0.5 °C compared to that in Fig. 4.10a. However, the ramp time of Fig. 4.10b during the recovery period is 3.09 min (= 12.08 min − 8.99 min) longer than that in Fig. 4.10a. This is because ACs in Fig. 4.10b have just reached the new temperature hysteresis band when they are recovered to the initial states and therefore it takes longer to migrate to the initial temperature hysteresis band according to the rule of SP-2 [35].

On the other hand, DT^* in Fig. 4.10a and c is the same (0.5 h), while RC^* of the latter is 9 MW higher than the former. It can be seen from Table 4.4 that ACs are divided into two groups to be dispatched and five group to be recovered both in

Fig. 4.10a and c. Table 4.3 shows that such increase of RC^* leads to the increase of *SD* from 0.30 MW to 0.83 MW, while *PV* is similar. All of the controllable ACs are utilized for the provision of operating reserve in Fig. 4.10c–e, where the percentage of power reduction are 63.6%, 80% and 100%, respectively. DT^* in Fig. 4.10c–e is set as the maximum deployment duration obtained from Fig. 4.8. Similar to Fig. 4.10b, ACs are divided into less group to be recovered during the recovery period in Fig. 4.10e, leading to the increase of *PV* by 1.75% (= 5.71% − 3.96%) than that in Fig. 4.10c and 1.52% (= 5.71% − 4.19%) than that in Fig. 4.10d. Hence, shorter duration time and larger reserve capacity will lead to larger fluctuations of aggregate power. As a result, the possible reserve capacity and duration time should be constrained by the maximum power fluctuations, which is quantified by *SD* and *PV*. Moreover, Table 4.3 shows that the ramp time of reserve deployment is within 5.98 min, which can fulfill the requirement on ramp time of 10 min spinning reserve, 30 min spinning reserve, etc. [31]. The ramp time during the recovery period is within 12.19 min, which is also shorter than the maximum limit (between 15 and 90 min) for different types of operating reserve [58]. Therefore, the sequential dispatch strategy of ACs can mitigate both the lead rebound and lag rebound entirely, which enables flexible control of the duration time and reserve capacity to fulfill the requirements of different types of operating reserve.

4.5.3 Comparison of Different Dispatch Strategy of ACs for the Provision of Operating Reserve

In order to validate the necessity of mitigating both the lead rebound and lag rebound, the operating reserve provided by ACs is utilized to relieve congestion resulted from the system peak load in IEEE-30-bus test system [59], diagram of which is shown in Fig. 4.12. To obtain the load patterns for the summer day, 1.23% of the historical hourly load in the COAST weather zone in Electric Reliability Council of Texas (ERCOT) in 24th August 2017 is utilized to generate the total system load [60]. In this way, the peak of total system load equals to approximately 120% of the load demand in standard IEEE-30-bus test system [59]. The profile of total system load [60] and the corresponding ambient temperature [61] are shown in Fig. 4.13.

Figure 4.13 shows that the system peak load exists at around 15:00. At 16:00, ACs located from bus 14–24 provide operating reserve of 14 MW as illustrated by Fig. 4.10c, which accounts for 15% of the electricity consumption in these buses. The operating reserve provided by ACs is replaced by 30 min operating reserve at 16:30, after which ACs are recovered to the initial states. The total number of controllable ACs located from bus 14 to bus 24 is generated from the consumer travel habit data collected by National Household Travel Survey [62] and is shown by the bars in Fig. 4.14. The total aggregate power corresponding to the controllable ACs is shown by the square-scattered lines in Fig. 4.14. The distribution of controllable ACs located

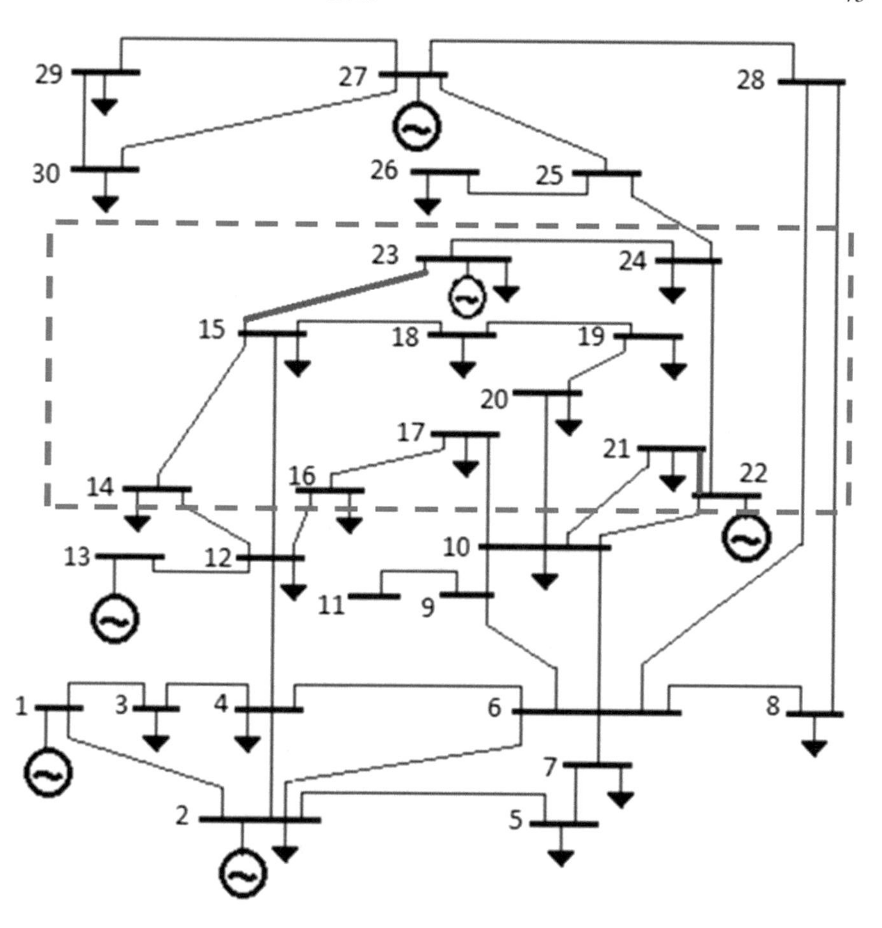

Fig. 4.12 Diagram of IEEE-30 bus system

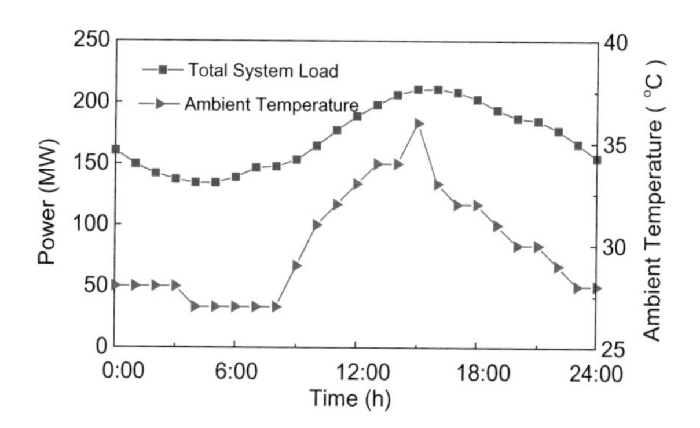

Fig. 4.13 Total system load and ambient temperature at each time instant

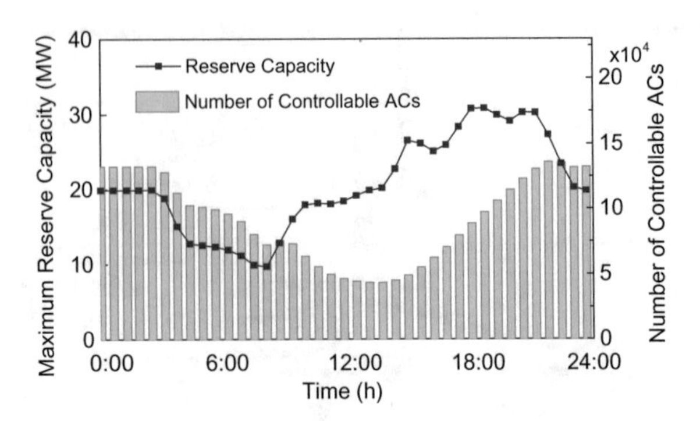

Fig. 4.14 Number of controllable ACs and the potential reserve capacity at each time instant

from bus 14 to bus 24 is proportional to the base load in these buses [59]. Locational marginal price (LMP) at each bus is evaluated by optimal power flow (OPF).

The proposed sequential dispatch strategy (SDS) is compared with the four other methods, which includes: (1) GDS [17, 27]: The concept of grouping devices to reduce the lag rebound in existing literatures, which divide ACs into several groups and recover them at a regular time interval. In this case, ACs are divided into three groups to be dispatched every 10 min and divided into five groups to be recovered every 10 min. In addition, the control of ACs in each group also follows safe protocol-2, which is the same as the proposed SDS; (2) RDS [13]: Randomizing the deployment/recovery of ACs over time, which is the most common way to mitigate the demand response rebound in existing researches. In this case, the deployment and recovery of ACs are randomized between 0 and 10 min; (3) SP-2 [46]: The safe protocol-2 to avoid synchronization of ACs, which reduces the power fluctuations and the level of demand response rebound; (4) CDS [63]: Traditional centralized dispatch strategy in which all the ACs are deployed/recovered when receiving reserve deployment/recovery signal instantly.

The total system load profile after the reserve deployment of ACs controlled by different dispatch strategies at 16:00 is shown in Fig. 4.15. The indices for the simulated cases are presented in Table 4.5. Branch *m-n* denotes the branch between bus *m* and bus *n*. The branch loading index (BLI) in branch 21–22 and branch 15–23 are shown by curves in Fig. 4.16. LMP in 5 min intervals in bus 21 is shown by the bars in Fig. 4.16.

The profile of BLI in Fig. 4.16 shows that congestion exists in branch 21–22 and branch 15–23 since 15:00 because of the system peak load, leading to the increase of LMP from 44$/MW to 72$/MW. The deployment of operating reserve provided by ACs at 16:00 relieves the congestion and reduces the LMP from 72$/MW to 44$/MW. Table 4.5 and the deployment segment from 16:00 to 16:30 in Fig. 4.15 show that the lead rebound is mitigated entirely in SDS. Accordingly, LMP in Fig. 4.16a remains at the level of 44$/MW after 16:00.

Similar to SDS, GDS can also mitigate the lead rebound entirely. However, the ramp time prolongs to 23.40 min, which cannot fulfill the requirement of many types

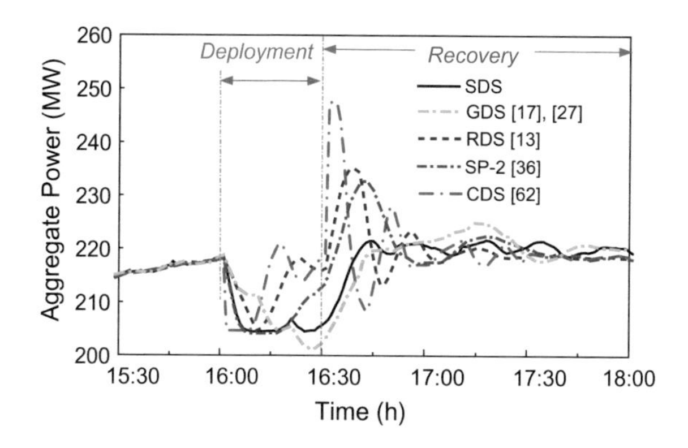

Fig. 4.15 Total system load after the reserve deployment of ACs controlled by different dispatch strategies at 16:00

Table 4.5 Indices of the operating reserve provided by ACs with different dispatch strategies

Dispatch strategies	DT (min)	BC_d (MW)	BC_r (MW)	RT_d (min)	RT_r (min)
Proposed SDS	30.00	0	0	5.98	8.52
GDS [17, 27]	30.00	0	5.22	23.40	8.68
RDS [13]	15.11	12.02	16.71	9.73	7.03
SP-2 [46]	19.27	6.13	17.89	4.87	8.95
CDS [63]	8.93	15.63	30.45	0.62	0.55

of operating reserves (e.g., 10 min spinning reserve). By contrast, RDS, SP-2 and CDS cannot mitigate the lead rebound entirely and the value reaches 12.02 MW, 6.13 MW and 15.63 MW, respectively. Compared to CDS and RDS, almost all the power fluctuations are removed by SP-2. However, SP-2 cannot entirely mitigate the lead rebound either and the aggregate power still rebound within the required duration time. Consequently, the actual duration time is only 19.27 min, which is shorter than DT^* (30 min). Congestion still exists during the reserve deployment period and the LMP increases to the level of 72$/MW again, as shown in Fig. 4.16c–e. Therefore, it is crucial to mitigate the lead rebound entirely so that the duration time can be flexibly controlled.

On the other hand, Table 4.5 and the recovery segment from 16:30 to 18:00 in Fig. 4.15 show that the lag rebound is also mitigated entirely by SDS, but cannot be mitigated entirely by RDS, SP-2 and CDS, whose lag rebounds reach 16.71 MW, 17.89 MW and 30.45 MW, respectively. As a result, the aggregate power attained by RDS, SP-2 and CDS are very large, resulting in the increase of LMP to over 72$/MW. Because of lacking the co-optimization among the reserve capacity and dispatch time instant of different AC groups in GDS, a rebound peak of 5.22 MW still exists at around 17:10, leading to the increase of LMP to around 67$/MW. It is required to mention that although the ramp time in SDS and SP-2 is a little longer than that in

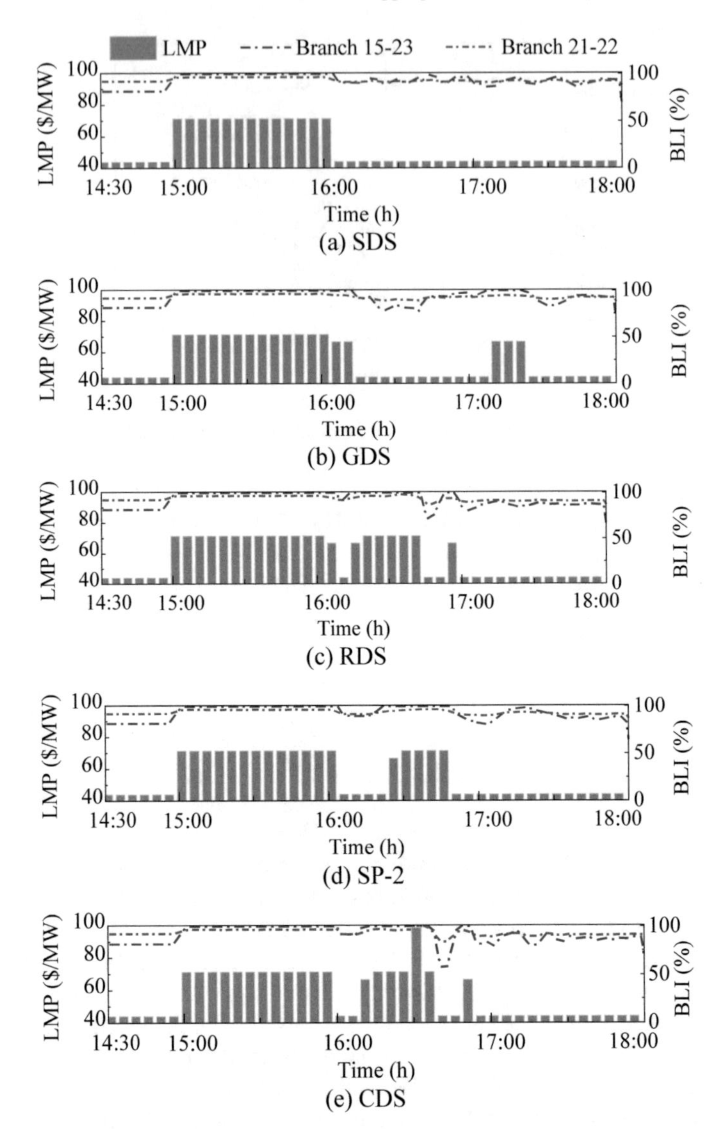

Fig. 4.16 Locational marginal price in 5-min intervals in bus 21 and branch loading index of the congestion branches corresponding to different control strategies

CDS and RDS, it can fulfill the requirements of most types of operating reserve. Therefore, the proposed SDS is better than the other dispatch strategies on the entire mitigation of the lead-lag rebound attained by the capacity-time co-optimization during the reserve deployment/recovery process.

4.6 Conclusions

This chapter presents a novel sequential dispatch strategy of ACs for the provision of the operating reserve. The impacts of the lead-lag rebound on the capacity dimension and the time dimension are quantified by a proposed evaluation framework. Illustrative results demonstrate that the sequential dispatch strategy and recovery algorithm enable ACs to provide operating reserve with multiple duration time. The maximum reserve capacity and the corresponding feasible duration time range are constrained by parameters including the total number of ACs and the power volatility limit. Aggregators should carefully balance these constraints to determine the reserve capacity and duration time. Moreover, the comparison of the proposed strategy with other methods illustrates that the proposed capacity-time co-optimization among different AC groups enable the entire mitigation of the lead-lag rebound. In this way, ACs can be utilized to relieve congestion or reduce peak load without adding additional burden to the power system.

References

1. F. Teng, V. Trovato, G. Strbac, Stochastic scheduling with inertia-dependent fast frequency response requirements. IEEE Trans. Power Syst. **31**(2), 1557–1566 (2016)
2. N. Zhang, C. Kang, Q. Xia, J. Liang, Modeling conditional forecast error for wind power in generation scheduling. IEEE Trans. Power Syst. **29**(3), 1316–1324 (2014)
3. H. Zhong, Q. Xia, C. Kang, M. Ding, J. Yao, S. Yang, An efficient decomposition method for the integrated dispatch of generation and load. IEEE Trans. Power Syst. **30**(6), 2923–2933 (2015)
4. C. Gao, Q. Li, Y. Li, Bi-level optimal dispatch and control strategy for air-conditioning load based on direct load control. Proc. CSEE **34**, 1546–1555 (2014)
5. M. Izquierdo, A. Moreno-Rodríguez, A. González-Gil, N. García-Hernando, Air conditioning in the region of Madrid, Spain: an approach to electricity consumption, economics and CO2 emissions. Energy **36**(3), 1630–1639 (2011)
6. S. Cox, Cooling a warming planet: a global air conditioning surge. https://e360.yale.edu/features/cooling_a_warm-ing_planet_a_global_air_conditioning_surge
7. P. Siano, C. Cecati, H. Yu, J. Kolbusz, Real time operation of smart grids via FCN networks and optimal power flow. IEEE Trans. Ind. Inform. **8**(4), 944–952 (2012)
8. C. Ju, P. Wang, L. Goel, and Y. Xu, A two-layer energy management system for microgrids with hybrid energy storage considering degradation costs. IEEE Trans. Smart Grid **99**, 1–1 (2017)
9. T. Jin, C. Kang, H. Chen, *Integrating Consumer Advance Demand Data in Smart Grid Energy Supply Chain. In Smart Grids: Clouds, Communications, Open Source, and Automation* (CRC Press, Boca Raton, FL, USA, 2014), pp. 251–274

10. V. Trovato, F. Teng, and G. Strbac, Role and benefits of flexible thermostatically controlled loads in future low-carbon systems. IEEE Trans. Smart Grid **99**, 1–1 (2017)
11. C.H. Wai, M. Beaudin, H. Zareipour, A. Schellenberg, N. Lu, Cooling devices in demand response: a comparison of control methods. IEEE Trans. Smart Grid **6**(1), 249–260 (2015)
12. K.P. Schneider, E. Sortomme, S.S. Venkata, M.T. Miller, L. Ponder, Evaluating the magnitude and duration of cold load pick-up on residential distribution using multi-state load models. IEEE Trans. Power Syst. **31**(5), 3765–3774 (2016)
13. A. Abiri-Jahromi, F. Bouffard, Contingency-type reserve leveraged through aggregated thermostatically-controlled loads-part I: characterization and control. IEEE Trans. Power Syst. **31**(3), 1972–1980 (2016)
14. N. Lu, S. Katipamula, Control strategies of thermostatically controlled appliances in a competitive electricity market. In Proceeding of IEEE Power Engineering Society General Meeting (San Francisco, CA, USA 2005), pp. 202–207
15. N.G. Paterakis, M. Gibescu, A.G. Bakirtzis, J.P.S. Catalão, A multi-objective optimization approach to risk-constrained energy and reserve procurement using demand response. IEEE Trans. Power Syst. **99**, 1–1 (2017)
16. D.T. Nguyen, M. Negnevitsky, M. de Groot, Modeling load recovery impact for demand response applications. IEEE Trans. Power Syst. **28**(2), 1216–1225 (2013)
17. N. Saker, M. Petit, J.L. Coullon, Demand side management of electrical water heaters and evaluation of the cold load pick-up characteristics (CLPU). In 2011 IEEE Trondheim PowerTech (2011), pp. 1–8
18. Y.W. Law, T. Alpcan, V.C.S. Lee, A. Lo, S. Marusic, M. Palaniswami, Demand response architectures and load management algorithms for energy-efficient power grids: a survey. In Proceedings of International Conference Knowledge Information Creativity Support Systems (Melbourne, Vic., Australia, 2012), pp. 134–141
19. T. Ericson, Direct load control of residential water heaters. Energy Policy **37**(9), 3502–3512 (2009)
20. P. Siano, D. Sarno, Assessing the benefits of residential demand response in a real time distribution energy market. Appl. Energy **161**, 533–551 (2016)
21. Q. Wu, P. Wang, L. Goel, Direct load control (DLC) considering nodal interrupted energy assessment rate (NIEAR) in restructured power systems. IEEE Trans. Power Syst. **25**(3), 1449–1456 (2010)
22. L. Goel, Q. Wu, P. Wang, Fuzzy logic-based direct load control of air conditioning loads considering nodal reliability characteristics in restructured power systems. Electr. Power Syst. Res. **80**(1), 98–107 (2010)
23. C.M. Chu, T.L. Jong, A novel direct air-conditioning load control method. IEEE Trans. Power Syst. **23**(3), 1356–1363 (2008)
24. B.J. Kirby, Load response fundamentally matches power system reliability requirements. In Proceeding of IEEE Power Engineering Society General Meeting (2007), pp. 1–6
25. P. Grünewald, J. Torriti, Demand response from the non-domestic sector: early UK experiences and future opportunities. Energy Policy **61**, 423–429 (2013)
26. N. Motegi, M.A. Piette, D.S. Watson, S. Kiliccote, P. Xu, *Introduction to Commercial Building Control Strategies and Techniques for Demand Response Appendices* (Lawrence Berkeley National Laboratory, USA, 2007)
27. H. Johal, K. Anaparthi, J. Black, Demand response as a strategy to support grid operation in different time scales. In 2012 IEEE Energy Conversion Congress and Exposition (ECCE) (2012), pp. 1461–1467
28. B.M. Sanandaji, T.L. Vincent, K. Poolla, Ramping rate flexibility of residential HVAC loads. IEEE Trans. Sustain. Energy **7**(2), 865–874 (2016)
29. X. Wu, J. He, Y. Xu, J. Lu, N. Lu, X. Wang, Hierarchical control of residential HVAC units for primary frequency control. IEEE Trans. Smart Grid **99**, 1–1 (2017)
30. N. Lu, An evaluation of the HVAC load potential for providing load balancing service. IEEE Trans. Smart Grid **3**(3), 1263–1270 (2012)

31. J. Wang, M. Shahidehpour, Z. Li, Contingency-constrained reserve requirements in joint energy and ancillary services auction. IEEE Trans. Power Syst. **24**(3), 1457–1468 (2009)
32. J.F. Ellison, L.S. Tesfatsion, V.W. Loose, R.H. Byrne, *Project Report: A Survey of Operating Reserve Markets in U.S. ISO/RTO-Managed Electric Energy Regions* (Sandia National Laboratories, USA, 2012)
33. W. Cui, Y. Ding, H. Hui, Z. Lin, P. Du, Y. Song, C. Shao, Evaluation and Sequential-Dispatch of Operating Reserve Provided by Air Conditioners Considering Lead-Lag Rebound Effect. IEEE Trans. Power Syst. (2018).
34. D.S. Callaway, Tapping the energy storage potential in electric loads to deliver load following and regulation, with application to wind energy. Energy Convers. Manag. **50**(5), 1389–1400 (2009)
35. C. Shao, Y. Ding, J. Wang, Y. Song, Modeling and integration of flexible demand in heat and electricity integrated energy system. IEEE Trans. Sustain. Energy **9**(1), 361–370 (2018)
36. D. Xie, H. Hui, Y. Ding, Z. Lin, Operating reserve capacity evaluation of aggregated heterogeneous TCLs with price signals. Appl. Energy **216**, 338–347 (2018)
37. A. Zipperer et al., Electric energy management in the smart home: perspectives on enabling technologies and consumer behavior. Proc. IEEE **101**(11), 2397–2408 (2013)
38. Q. Hu, F. Li, X. Fang, L. Bai, A framework of residential demand aggregation with financial incentives. IEEE Trans. Smart Grid **9**(1), 497–505 (2018)
39. Q. Wang, C. Zhang, Y. Ding, G. Xydis, J. Wang, J. Østergaard, Review of real-time electricity markets for integrating distributed energy resources and demand response. Appl. Energy **138**, 695–706 (2015)
40. J. Wang, C.N. Bloyd, Z. Hu, Z. Tan, Demand response in China. Energy **35**(4), 1592–1597 (2010)
41. M. Parvania, M. Fotuhi-Firuzabad, Integrating load reduction into wholesale energy market with application to wind power integration. IEEE Syst. J. **6**(1), 35–45 (2012)
42. M. Shafie-khah, P. Siano, A stochastic home energy management system considering satisfaction cost and response fatigue. IEEE Trans. Ind. Informat. **99**, 1–1 (2017)
43. Q. Hu, F. Li, Hardware design of smart home energy management system with dynamic price response. IEEE Trans. Smart Grid **4**(4), 1878–1887 (2013)
44. C. Perfumo, E. Kofman, J.H. Braslavsky, J.K. Ward, Load management: Model-based control of aggregate power for populations of thermostatically controlled loads. Energy Convers. Manag. **55**, 36–48 (2012)
45. J.L. Mathieu, S. Koch, D.S. Callaway, State estimation and control of electric loads to manage real-time energy imbalance. IEEE Trans. Power Syst. **28**(1), 430–440 (2013)
46. N.A. Sinitsyn, S. Kundu, S. Backhaus, Safe protocols for generating power pulses with heterogeneous populations of thermostatically controlled loads. Energy Convers. Manag. **67**, 297–308 (2013)
47. N. Mehta, N.A. Sinitsyn, S. Backhaus, B.C. Lesieutre, Safe control of thermostatically controlled loads with installed timers for demand side management. Energy Convers. Manag. **86**, 784–791 (2014)
48. B. Zhao, J. Liu, G. Zhang, J. Su, Recovery strategy and probability model for emergency control of thermostatically controlled load groups. In 2017 IEEE Transportation Electrification Conference and Expo, (ITEC Asia-Pacific), (Asia-Pacific 2017), pp. 1–5
49. J. Bendtsen, S. Sridharan, Efficient desynchronization of thermostatically controlled loads. IFAC Proc. **46**(11), 245–250 (2013)
50. E.G. Talbi, *Metaheuristics for Bi-Level Optimization* (Springer, Berlin Heidelberg, 2013)
51. J.L. Mathieu, M. Dyson, D.S. Callaway, Using residential electric loads for fast demand response: the potential resource and revenues, the costs, and policy recommendations. In Proceedings of the ACEEE Summer Study on Buildings (2012)
52. U. E. I. Administration, Residential energy consumption survey (RECS), (2012). https://www.eia.gov/todayinenergy
53. Z. Jean-Paul, E. Matt, G. Jonathan, K. Nicola, Household electricity survey. Department for Environment Food and Rural Affairs (Defra), U.K. (2012)

54. G. of S. Australia, Home energy use. https://www.sa.gov.au/topics/energy-and-environment/using-saving-energy/home-energy-use
55. X. Zheng, C. Wei, P. Qin, J. Guo, Y. Yu, F. Song, Z. Chen, Characteristics of residential energy consumption in China: findings from a household survey. Energy Policy **75**, 126–135 (2014)
56. B.J. Kirby, *Spinning Reserves from Controllable Packaged Through the Wall Air Conditioner (PTAC) Units* (Oak Ridge National Laboratory, USA, 2003)
57. H. Hui, Y. Ding, W. Liu, Y. Lin, Y. Song, Operating reserve evaluation of aggregated air conditioners. Appl. Energy **196**, 218–228 (2017)
58. E. Ela, M. Milligan, B. Kirby, *Operating Reserves And Variable Generation* (National Renewable Energy Laboratory, USA, 2011)
59. M. Shahidehpour, Y. Wang, Appendix C: IEEE30 bus system data. In Communication and Control in Electric Power Systems: Applications of Parallel and Distributed Processing, (Wiley-IEEE Press, 2003)
60. Electric Reliability Council of Texas, Hourly load data archives (2017). http://www.ercot.com/gridinfo/load/load_hist
61. Time and Date AS, Weather in August 2017 in Houston, Texas, USA. https://www.timeanddate.com/weather/usa/hous-ton/historic?month=8&year=2017
62. U.S. Department of Transportation, Federal Highway Administration, National household travel survey (2009). http://nhts.ornl.gov
63. C. Perfumo, J. Braslavsky, J.K. Ward, A sensitivity analysis of the dynamics of a population of thermostatically-controlled loads. In 2013 Australasian Universities Power Engineering Conference (AUPEC), pp. 1–6 (2013)

Chapter 5
Inverter Air Conditioner Aggregation for Providing Frequency Regulation Service

5.1 Introduction

The above four chapters focus on the operating reserve provided by ACs or TCLs, in which frequency regulation service (FRS) is one of the most important operating reserve for the power systems. The high penetration of intermittent renewables and the increasing severe contingencies of generation, transmission, and distribution infrastructures have led to more fluctuations faced by current power systems [1]. For example, the bipolar locking of the ultra-high voltage direct current transmission line in the East China Grid resulted in the system frequency decline by 0.41 Hz abruptly on September 9th, 2015 [2], and the large-scale blackout in Taiwan impacted 5.92 million customers on August 15th, 2017 [3]. Therefore, the FRS is becoming increasingly important by maintaining power balance between generation and consumption. Conventionally, FRS is provided by generators, such as thermal power generators [4]. However, traditional generators may be phased out in the future due to the constraint of the greenhouse gas emissions on a global scale, making the traditional generators insufficient to deal with the increasing requirements of FRS [5]. The information and communication technology has improved a lot over the past decades, and in the meantime smart home appliances become more popular, which make it easier for loads to be controlled directly to assist the system in maintaining balance [6, 7]. Loads can reduce/increase operating power to provide FRS when the system frequency drops/rises [8]. Moreover, the operating power of demand side resources (DSRs) can be regulated rapidly, while the generator regulates its power generation through a series of processes, such as the speed governor process and the reheat steam turbine process, leading to a larger inertia compared with DSRs [9]. Meanwhile, customers can get benefits for their contributions to maintaining the system stability [10, 11]. Therefore, several studies are turning from supply side to demand side [12, 13]. For example, Benysek et al. [14] develops an application-ready control algorithm based on the stochastic and decentralized strategy to realize the load frequency control. A hybrid hierarchical control scheme of demand side resources is developed to support FRS in [15, 16].

Y. Ding et al., *Integration of Air Conditioning and Heating into Modern Power Systems*,
https://doi.org/10.1007/978-981-13-6420-4_5

Among common home appliances, such as lights, televisions, air conditioners (ACs), refrigerators and water heaters, the ACs top the list of power consumption [17, 18]. A short time regulation of the operating power of the AC has little effect on the room temperature due to the heat preservation property [5]. Therefore, ACs are suitable and have huge potential to serve as DSR. Apart from regular fixed speed ACs, the market share of inverter ACs is expanding rapidly [19]. For example, the sale volume of inverter ACs in China has exceeded the regular fixed speed ACs [20]. The main difference of the regular and inverter ACs is the compressor, which is also the main power consumption component of the AC [21]. The regular AC's compressor operates in only two modes, i.e., on- or off-mode. Therefore, the operating power of regular ACs can be approximated as switching between the rated power and zero. In contrast, the compressor's speed of the inverter AC can be adjusted continuously by changing the operating frequency, making it more flexible to adjust AC's operating power and follow FRS instructions [21].

References [22, 23] develop an inverter AC model based on the simulation method and experimental data, respectively. However, these two studies only focus on the physical model and do not consider the interaction between the ACs and power systems. References [24, 25] present an accurate mathematical model of ACs to provide FRS for power systems, while the model is based on regular ACs with fixed speeds. Besides, a control method considering customer's set temperature is proposed in [26] to dispatch regular ACs for providing balancing services for power systems, while this method is developed particularly for the AC and the system operator has to dispatch the generators and the ACs in two different ways.

In existing power systems, the FRS is mainly provided by traditional generators, such as thermal power generators [4, 9]. Many sophisticated control methods for these traditional generators have been developed [27]. For example, generators are equipped with speed governors to provide primary frequency regulation (PFR) with the proportional control method. Some generators are installed with synchronizers to participate in secondary frequency regulation (SFR) with the integral control method [28]. However, the operation characteristics of the inverter AC are different from traditional generators. A major difference rests in that the inverter AC is not installed with speed governors or synchronizers. Some control methods for the inverter AC have been proposed such as changing the operating states (on/off control method) [6] or adjusting the set temperatures to achieve adjustable operating power [5]. Although the above control methods for the generator and inverter AC have been developed in previous studies, the dispatching models of the generator and inverter AC are remarkably distinct. It is not yet clear how to dispatch generators and inverter ACs to provide FRS simultaneously by using the same set of the control system.

To address this issue, this chapter develops a novel thermal and electrical model of the inverter AC, which is equivalent to a generator, so that the inverter AC can be controlled as a generator to provide FRS. In this manner, the existing control system for generators can send scheduling instructions to inverter ACs (similar as those sent to generators), making it more accessible for inverter ACs to participate in FRS. The main contributions of this chapter are as follows:

(a) A novel thermal and electrical model of the inverter AC for providing FRS is developed. Based on this model, the inverter AC can be controlled to change the operating power for providing FRS.
(b) The model of the inverter AC is derived and equivalent to a traditional generator, including the control parameters and evaluation criteria. In this manner, the inverter AC can be scheduled and compatible with the existing control system.
(c) A stochastic allocation method of the regulation sequence among inverter ACs is proposed to reduce the effect of FRS on customers. Besides, a hybrid control strategy by taking into account the dead band control and the hysteresis control is developed to reduce the frequency fluctuations of power systems.

This chapter includes research related to the equivalent modeling of inverter air conditioners for providing frequency regulation service by [29].

5.2 Thermal and Electrical Model of the Inverter AC Considering Providing FRS

5.2.1 Thermal Model of a Room

To study the operating characteristics of inverter ACs, it is necessary to develop the thermal model of a room. Some valid models have been built to describe the relationship between the room temperature and the thermal deviation [5, 22], which can be expressed as [26]:

$$c_A \rho_A V \cdot \Delta T_A = \int (\Delta Q_{gain} - \Delta Q_{AC}) dt \tag{5.1}$$

$$\Delta Q_{gain} = (U_{O-A} A_S + c_A \rho_A V \xi)(\Delta T_O - \Delta T_A) + \Delta Q_{dis} \tag{5.2}$$

where Δ denotes the deviation of the parameters; c_A is the heat capacity of the air; ρ_A is the density of the air; V and A_S are the volume and the surface area of the room, respectively; T_A is the indoor temperature; Q_{gain} is the total heat gains of the room; Q_{AC} is the refrigerating capacity of the inverter AC; U_{O-A} and ξ are the heat transfer coefficient and air exchange times between the room and the ambience, respectively; T_O is the ambient temperature; Q_{dis} is the heat power from people, lights, appliances and other disturbances.

The thermal model can also be expressed in the frequency domain by the Laplace Transform:

$$c_A \rho_A V \cdot T_A(s)s = Q_{gain}(s) - Q_{AC}(s) \tag{5.3}$$

$$Q_{gain}(s) = (U_{O-A}A_S + c_A\rho_A V\xi)[T_O(s) - T_A(s)] + Q_{dis}(s) \tag{5.4}$$

where s is the Laplace operator.

5.2.2 Electrical Model of an Inverter AC Considering Providing FRS for Power Systems

The major differences between inverter ACs and regular fixed speed ACs are the frequency converter and the compressor. The regular AC's compressor only works in a fixed speed, while the inverter AC's compressor can change the speed continuously by adjusting the operating frequency. The operating power and refrigerating capacity are also regulated with the operating frequency, which can be expressed as:

$$\Delta P_{AC} = \kappa_P \Delta f_{AC}\left(1 - e^{-t/T_c}\right) \tag{5.5}$$

$$\Delta Q_{AC} = \kappa_Q \Delta f_{AC}\left(1 - e^{-t/T_c}\right) \tag{5.6}$$

where P_{AC} and Q_{AC} are the operating power and refrigerating capacity of the inverter AC, respectively; f_{AC} is the operating frequency of the inverter AC; κ_P and κ_Q are the constant coefficients of the inverter AC; T_c is the time constant of the compressor. The electrical model can also be expressed in the frequency domain [22]:

$$P_{AC}(s) = \frac{\kappa_P}{T_c s + 1} f_{AC}(s) + \mu_P \tag{5.7}$$

$$Q_{AC}(s) = \frac{\kappa_Q}{T_c s + 1} f_{AC}(s) + \mu_Q \tag{5.8}$$

where μ_P and μ_Q are the constant coefficients of the inverter AC. Therefore, the relationship between the operating power and the refrigerating capacity can be described as:

$$Q_{AC}(s) = \frac{\kappa_Q}{\kappa_P} P_{AC}(s) + \frac{\kappa_P \mu_Q - \kappa_Q \mu_P}{\kappa_P} \tag{5.9}$$

The operating frequency of the inverter AC is mainly based on the gap between the set temperature and the current room temperature, which can be expressed as:

$$\Delta f_{AC}(s) = C(s) \cdot \Delta T_{dev}(s) \tag{5.10}$$

$$\Delta T_{dev}(s) = \Delta T_A(s) - \Delta T_{set}(s) \tag{5.11}$$

where $C(s)$ is the temperature controller of the inverter AC; T_{dev} is the deviation between the indoor temperature T_A and the set temperature T_{set}.

Proportional integral (PI) controller is a conventional classical control method adopted by inverter ACs, which can meet the control requirement simply and effectively. The PI controller can be described as [30]:

$$C(s) = \theta + \eta / s \tag{5.12}$$

where θ and η are the constant coefficients of the controller.

In the time domain, the operating frequency of the inverter AC can be expressed as:

$$\Delta f_{AC} = \theta \cdot \Delta T_{dev} + \eta \cdot \int \Delta T_{dev} dt \tag{5.13}$$

If the inverter AC provides FRS for power systems, the AC's operating frequency will also be influenced by the system frequency, which can be described as:

$$\Delta f_{AC}(s) = C(s) \cdot \Delta T_{dev}(s) + D(s) \cdot \Delta f(s) \tag{5.14}$$

where Δf is the frequency deviation of the power system; $D(s)$ is the controller of the inverter AC for participating in FRS.

In the time domain, the operating frequency can be expressed as:

$$\Delta f_{AC} = \theta \cdot \Delta T_{dev} + \eta \cdot \int \Delta T_{dev} dt + \delta \cdot \Delta f + \gamma \cdot \int \Delta f dt \tag{5.15}$$

From Eqs. (5.5) and (5.15), the operating power of the inverter AC can be expressed as:

$$\Delta P_{AC} = \kappa_P (1 - e^{-t/T_c})(\theta \cdot \Delta T_{dev} + \eta \cdot \int \Delta T_{dev} dt + \delta \cdot \Delta f + \gamma \cdot \int \Delta f dt) \tag{5.16}$$

5.2.3 Analysis of the Thermal and Electrical Model

The above thermal and electrical model can be derived as Fig. 5.1, where Γ_A and Ψ_A equal to $(c_A \rho_A V)$ and $(U_{O-A} A_S + c_A \rho_A V \xi)$, respectively. The model is divided into five portions. Portion (c) is the main body of this model and all the other four portions connect with it. The dead band of the temperature gap is set to prevent the operating frequency from adjusting too frequently under tiny perturbations. The rate limiter is set to limit the adjusting speed of the compressor to ensure its safety. Besides, the deviation range of the operating frequency is limited in the saturation function.

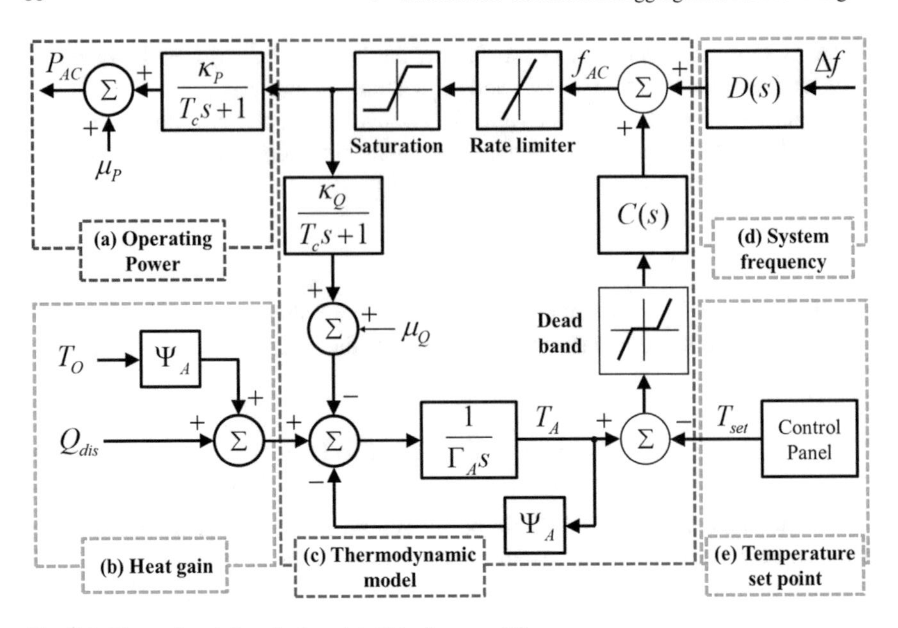

Fig. 5.1 Thermal and electrical model of the inverter AC

The Portion (b), (d) and (e) in this model can affect the operating power of the inverter AC. Portion (b) is the heat source of the model, which includes the transferred heat through the room envelope, the exchanged heat through the gaps of the doors or windows and the radiated heat from people or appliances. Portion (e) is the expected comfortable temperature of the customer, which can be adjusted through the remote control panel. Portion (d) is the additional signal for the inverter AC to participate in FRS. When the system frequency decreases, the inverter AC's operating frequency will also decrease to cut down the operating power and assist the recovery of the system frequency. Deviations in the above three portions will finally affect the operating power in Portion (a).

Figure 5.2 shows the analysis of the inverter AC's response speed in the time domain. It is assumed that the system frequency deviates from the rated value abruptly and the Δf is regarded as a step signal. The desired response speed is shown in Case 1. However, in fact, the adjusting speed of the compressor is limited by the rate limiter to ensure its safety, which is considered in Case 2. The ΔP_{AC} increases rapidly and reaches the maximum value within ten seconds.

As for the response time for PFR, the Union for the Co-ordination of Transmission of Electricity in Europe [31, 32] requires that the generator should reach the required value within thirty seconds. Moreover, SFR should start to respond within thirty seconds and reach the required value within fifteen minutes. Therefore, the FRS provided by the inverter AC can meet the requirements of the response time.

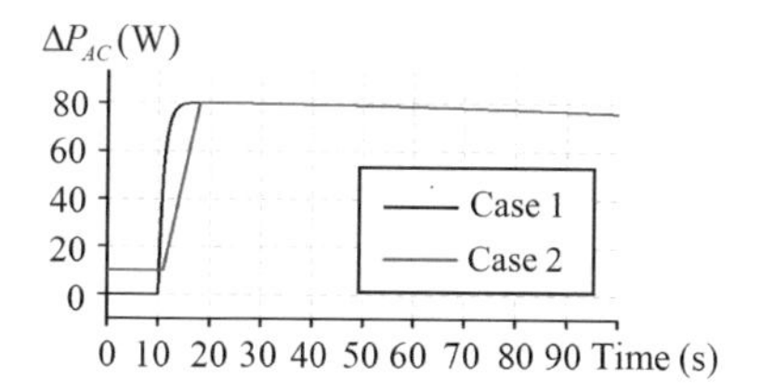

Fig. 5.2 Analysis of the inverter AC's response speed in the time domain

5.3 Equivalent Modeling of Inverter ACs for Providing Frequency Regulation Service

5.3.1 Equivalent Modeling of Inverter ACs

This chapter takes reheat steam generators as an example to analyze the characteristics of FRS [9]. The power-frequency regulation model of the reheat steam generator can be expressed as:

$$\Delta P_G(s) = -\frac{(1/R + K/s)(F_{HP}T_r s + 1)}{(T_g s + 1)(T_t s + 1)(T_r s + 1)}\Delta f(s) \tag{5.17}$$

where R is speed droop parameter of PFR; K is the integral gain of SFR; F_{HP} is the power fraction of the high pressure turbine section; T_g, T_r and T_t are the speed governor time constant, reheat time constant and turbine time constant, respectively.

From Eq. (5.1)–(5.16), the operating power deviation of the inverter AC can be derived as:

$$\Delta P_{AC}(s) = \frac{\kappa_P(\Gamma_A s + \Psi_A)(D(s)\Delta f(s) + C(s)\Delta T_{set}(s))}{(T_c s + 1)(\Gamma_A s + \Psi_A) + \kappa_Q C(s)} + \frac{\kappa_P C(s)(\Psi_A \Delta T_O(s) + \Delta Q_{dis}(s))}{(T_c s + 1)(\Gamma_A s + \Psi_A) + \kappa_Q C(s)} \tag{5.18}$$

where the operating power deviation ΔP_{AC} is affected by four factors: the system frequency deviation Δf, the set temperature deviation ΔT_{set}, the ambient temperature deviation ΔT_O and the radiated heat deviation ΔQ_{dis}. Generally, the duration time of the FRS process is short (within 30 s) [31, 32]. Therefore, the ambient temperature and the radiated heat can be assumed as invariable. When the inverter AC system enters a stable operating state, the set temperature is fixed, unless the customer adjusts the set value exactly at the time interval when the inverter AC is providing FRS. To sum up, considering the short FRS time, the three deviations ΔT_{set}, ΔT_O and ΔQ_{dis} can be omitted with regard to a stable operating inverter AC. Therefore, the ΔP_{AC} can be simplified to:

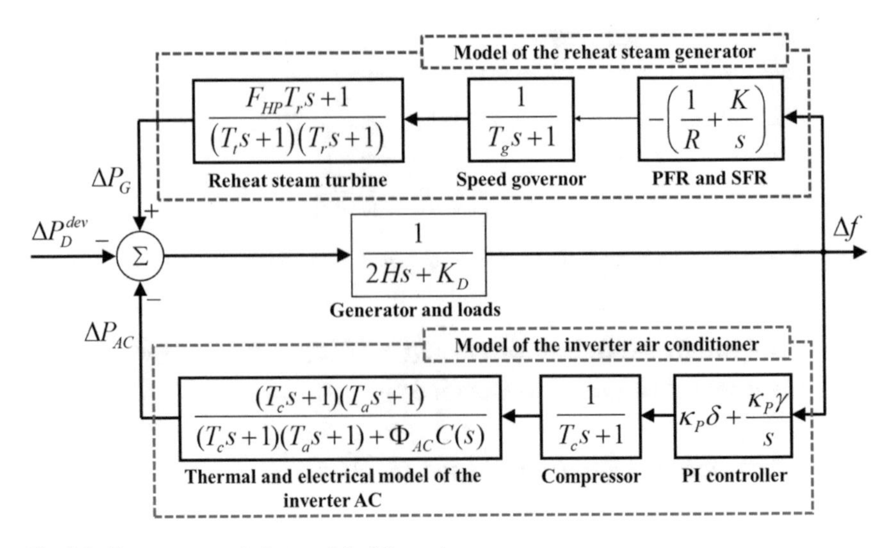

Fig. 5.3 Frequency regulation model of the system

$$\Delta P_{AC}(s) = \frac{\kappa_P D(s)(T_a s + 1)}{(T_c s + 1)(T_a s + 1) + \Phi_{AC} C(s)} \Delta f(s) \tag{5.19}$$

where T_a and Φ_{AC} equal to (Γ_A/Ψ_A) and (κ_Q/Ψ_A), respectively.

If the frequency regulation strategy of the inverter AC is the same with generators, the function $D(s)$ can be described as:

$$D(s)=\delta + \gamma / s \tag{5.20}$$

where δ and γ are the constant coefficients of the controller.

As shown in Fig. 5.3, the power generation of generators and the power consumption of inverter ACs both are related to the system frequency. When the system frequency decreases, the generator will increase power generation, while the inverter AC will decrease operating power. Both adjustments contribute to the stability of the system frequency.

Figure 5.4 shows the Bode diagram of the power system before and after considering the FRS provided by inverter ACs. The phase margin and gain margin are two important parameters of the stability criteria in the Bode diagram. In Case 1, the FRS is provided only by generators, where the phase margin and the gain margin are 51° and 27.9 dB, respectively. In Case 2, the FRS is provided by both generators and inverter ACs, where the phase margin widens to 80° and the gain margin is expanded towards infinity (Here the operating power of inverter ACs has no limitation). It indicates that the system's stability and frequency regulation ability get improved after the inverter ACs participating in FRS.

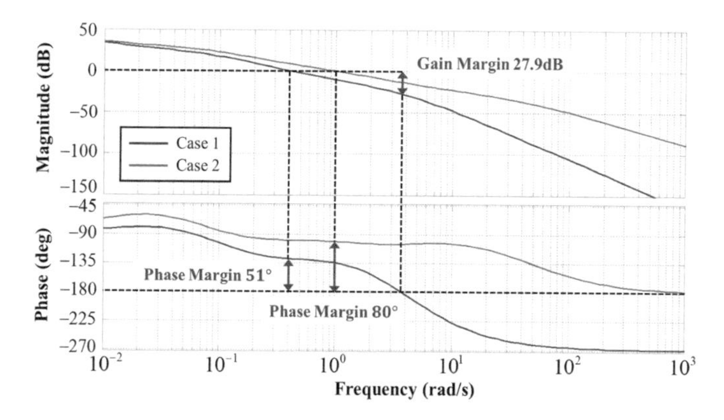

Fig. 5.4 Bode diagram for the frequency domain analysis of the power system considering the FRS provided by inverter ACs

5.3.2 *Equivalent Control Parameters*

(1) Equivalence of the PFR and SFR

Every generator can participate in PFR because all of them are equipped with speed governors [4]. The PFR is able to respond within 30 s [31] and mainly deals with random loads. The speed droop parameter of PFR is defined as R [9]. Correspondingly, the proportional controller ($\kappa_P \delta$) in the inverter AC model is the equivalent regulation parameter to provide PFR. However, the system frequency cannot recover the rated value only by PFR. Therefore, some generators are installed synchronizers to provide SFR. The integral coefficient of SFR is defined as K [4]. Correspondingly, the inverter AC can also provide SFR by the equivalent integral parameter $\kappa_P \gamma$.

(2) Equivalence of the First-Order Inertia Element

The first-order inertia element exists in the two models, which are the speed governor in the generator model and the compressor in the inverter AC model, respectively. Due to the little inertia of the compressor, the time constant T_c is less than the time constant of the speed governor T_g. Therefore, the inverter AC can regulate the operating power more rapidly than traditional generators.

(3) Equivalence of the Power Performing Component

The power performing component of the generator is the reheat steam turbine. Correspondingly, the inverter AC regulates the operating power by the thermal and electrical model, which can change the operating power temporarily without exceeding the customer's comfort temperature interval. The equivalent parameters are shown in Table 5.1.

Table 5.1 Equivalent parameters of generators and inverter ACs

Parameter	Generator	Inverter AC
Proportionality coefficient of PFR	$1/R$	$\kappa_P \delta$
Integral coefficient of SFR	K	$\kappa_P \gamma$
Time constant of the first-order inertia element	T_g	T_c
Power performing component	T_r, T_t, F_{HP}	T_c, T_a, Φ_{AC}
Regulation capacity	ΔP_{GP}	ΔP_{AC}
Activation time	AT_G	AT_{AC}
Delay time	DEL_G	DEL_{AC}

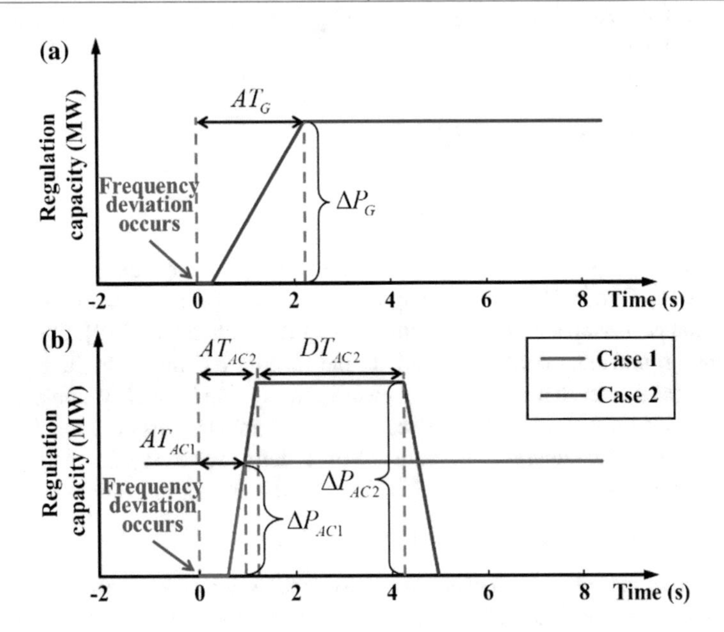

Fig. 5.5 Evaluation parameters for providing FRS: **a** the generator, **b** the inverter AC

5.3.3 *Equivalent Evaluation Parameters*

(1) **Regulation Capacity**

As shown in Fig. 5.5a, the generator will increase the power generation when the system frequency decreases. The maximum regulation capacity can be expressed as ΔP_G^{max}. Similarly, the inverter AC will provide regulation capacity by decreasing the operating power when the frequency deviation occurs. From Eq. (5.1)–(5.9), the maximum regulation capacity of the inverter AC can be expressed as:

$$\Delta P_{AC}^{max} = \kappa_P f_{AC}^{min} - \frac{\kappa_Q}{\kappa_P}\left[\Psi_A(T_O - T_A) + Q_{dis} - \mu_Q\right] \tag{5.21}$$

(2) **Activation Time**

The activation time is the time interval from the occurrence moment of the frequency deviation to the moment when the regulation capacity reaches the maximum value. The activation time is expressed as AT_G in the generator model and AT_{AC} in the inverter AC model. As shown in Fig. 5.5b, the activation time AT_{AC} and the regulation capacity ΔP_{AC} of the inverter AC are different in the two cases. Due to the limitation of the compressor's adjusting speed, a larger regulation capacity needs a longer activation time.

(3) **Duration Time**

The additional power generation of the generator can last more than 15 min until reserve generators are dispatched to make up the shortage of power [31, 32]. Therefore, the FRS provided by generators can be assumed to continue for an infinite time.

Different from generators, the FRS provided by inverter ACs may not continue for a long time. If the operating power is adjusted widely, the room temperature will change rapidly and reach the limitation of the comfortable temperature. Then the inverter AC has to increase operating power to ensure the customer's comfort. Therefore, the duration time DT_{AC}^{max} is defined to evaluate the relationship between the available regulation time and the regulation capacity ΔP_{AC}, which can be derived as:

$$DT_{AC}^{max} = \begin{cases} +\infty & ,0 \le |\Delta P_{AC}| \le \kappa_P \Delta T_A^{max} / \Phi_{AC} \\ -T_a \cdot \ln\left(1 + \dfrac{\kappa_P \Delta T_A^{max}}{\Phi_{AC}\Delta P_{AC}}\right) & ,\kappa_P \Delta T_A^{max} / \Phi_{AC} < |\Delta P_{AC}| \le \left|\Delta P_{AC}^{max}\right| \end{cases} \tag{5.22}$$

As shown in Fig. 5.6, the DT_{AC}^{max} will be shorter with the increase of ΔP_{AC}. By contrast, the DT_{AC}^{max} will be longer with the increase of the maximum allowable deviation of the room temperature ΔT_A^{max}.

5.4 Control of Aggregated Inverter ACs for Providing Frequency Regulation Service

5.4.1 The Regulation Capacity Allocation Among Generators and Inverter ACs

The regulation capacity of one inverter AC is paltry for the system. Therefore, the inverter ACs are aggregated and equivalent as a traditional generator. The maximum regulation capacity of the aggregation can be calculated as:

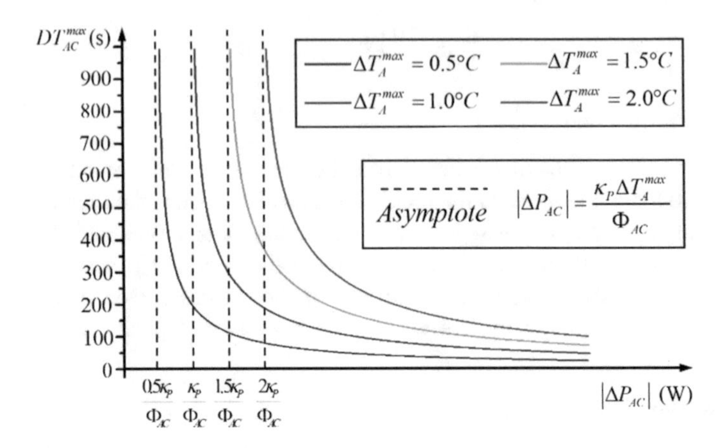

Fig. 5.6 The longest duration time of the inverter AC

$$\Delta P_{AC}^{max} = \sum_{i=1}^{N} \Delta P_{ACi}^{max} \tag{5.23}$$

Generally, the required regulation capacity of FRS (P_{RE}) is proportional to the loads in the system [4]. Each participant for FRS in the system will be allocated a certain proportion of the P_{RE}, which can be expressed as:

$$\begin{aligned} P_{RE} &= \alpha_G P_{RE} + \alpha_{AC} P_{RE} = \sum_{j=1}^{M} \alpha_{Gj} P_{RE} + \sum_{i=1}^{N} \alpha_{ACi} P_{RE} \\ &= \sum_{j=1}^{M} \Delta P_{Gj}^{max} + \sum_{i=1}^{N} \Delta P_{ACi}^{max} \end{aligned} \tag{5.24}$$

where α_G and α_{AC} are the shares of P_{RE} provided by generators and inverter ACs, respectively. M and N are the number of generators and inverter ACs, respectively.

Generators can provide FRS by increasing power generation and last until the reserve generators are dispatched to make up the power shortage. However, inverter ACs will return to the original operating power when the system frequency recovers to the rated value. Therefore, the inverter ACs can mainly provide PFR, while SFR is still mainly provided by the generators. In order to avoid the capacity shortage of SFR, the maximum share of inverter ACs in the total regulation capacity should be limited to a safety threshold, which can be expressed as:

$$\alpha_{AC} \leq \chi\% \tag{5.25}$$

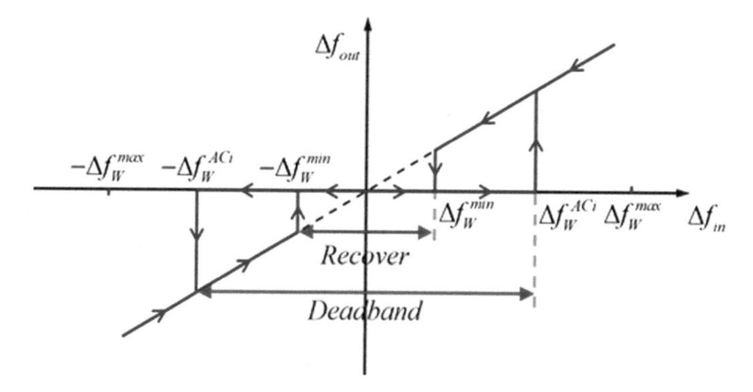

Fig. 5.7 The hybrid control strategy of the inverter AC for providing FRS

5.4.2 The Control Strategy of Inverter ACs

With increasing power system's frequency deviation, more inverter ACs should participate in FRS to provide more regulation capacity. The threshold value Δf_W^{ACi} is defined as the frequency deviation where the i-th inverter AC starts to provide FRS. Each inverter AC's Δf_W^{ACi} is uniformly distributed in the range from Δf_W^{min} to Δf_W^{max}, which produces the sequence of inverter ACs to provide FRS. It is obvious that no inverter AC provides FRS if the system frequency deviation is tiny and less than Δf_W^{min}, while all the inverter ACs will provide FRS if the system frequency deviation is larger than Δf_W^{max}.

As for an individual inverter AC, if the Δf_W^{ACi} is small, it will be in the front of the regulation sequence. In order to avoid a certain inverter AC always being at the forefront, each inverter AC's Δf_W^{ACi} is reset and generated randomly in each round of dispatch (every 15 min).

As shown in Fig. 5.7, the hybrid control strategy for inverter ACs takes into account the dead band control and the hysteresis control. The i-th inverter AC selected by the abovementioned allocation method will start participating in FRS when the Δf exceeds $\pm\Delta f_W^{ACi}$, while the inverter AC will withdraw from FRS when the Δf is returned to the range $\pm\Delta f_W^{min}$.

5.4.3 The Communication and Control Process of Inverter ACs

The communication of the system is shown in Fig. 5.8, where the signal sequence is labeled from one to five. First, before the next round of dispatch, the system operator sends the regulation capacity shares, i.e., α_G and α_{AC}, to the generators and aggregated inverter ACs, respectively. Secondly, the aggregator will communicate with each inverter AC to determine whether it is available for the FRS. The threshold

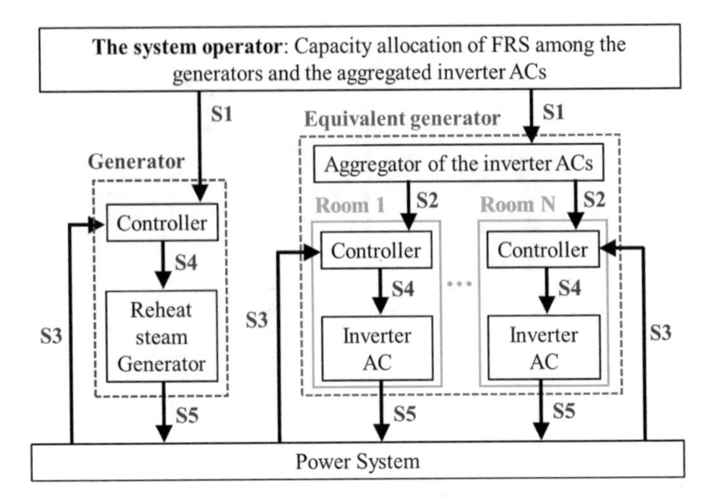

Fig. 5.8 The communication of the system

value of Δf_W^{ACi} will also be stochastically set in the controller between Δf_W^{min} and Δf_W^{max} for each available AC. Thirdly, after the new round of dispatch starts, the controllers of the generators and inverter ACs will monitor the system frequency locally in real time. Fourthly, if there is a frequency deviation, the controllers will send signals to the generators and inverter ACs to provide FRS. Finally, the power system's frequency will be regulated by the generators and inverter ACs.

As illustrated in Fig. 5.9, the control process of the inverter ACs can be divided into three steps.

The first step is the initialization of the system, where initial parameters of the network, the reheat steam generator, loads and inverter ACs are set. The share of the required regulation capacity is also allocated among generators and the aggregated inverter ACs.

The second step is the regulation capacity allocation among inverter ACs. The aggregator will communicate with each inverter AC to set the threshold value Δf_W^{ACi} for the available ACs. Besides, the aggregator will evaluate the maximum regulation capacity ΔP_{ACi}^{max} and keep communicating with more inverter ACs until the total regulation capacity reaches the required value.

The third step is the control of each inverter AC. The system frequency are detected locally. If the system frequency deviation exceeds the threshold Δf_W^{ACi} and the room temperature is within the maximum allowable deviation ΔT_{Ai}^{max}, the inverter AC will adjust its operating power to provide FRS. Otherwise, the inverter AC will keep operating at the normal state. After the third step ends, the program will repeat steps (1)–(3) for the next round of dispatch.

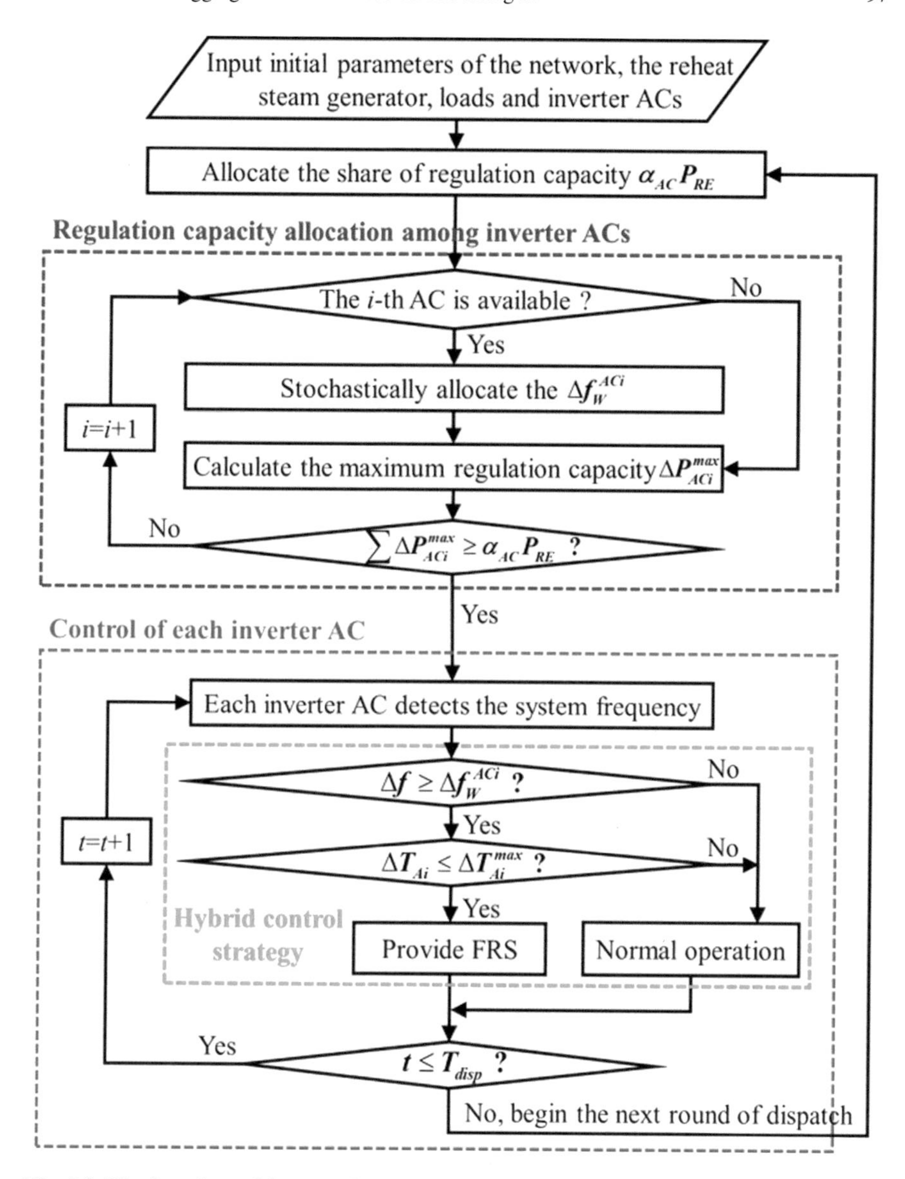

Fig. 5.9 The flow chart of the control process of inverter ACs

Table 5.2 Constant parameters of the thermal model and electrical model

Symbols	Descriptions	Values	Units
c_A	Heat capacity of the air	1.005	kJ/(kg Δ °C)
ρ_A	Density of the air	1.205	kg/m^3
ξ	Air exchange times	0.50	$1/h$
U_{O-A}	Heat transfer coefficient	3.60	W/m^2 Δ °C
Q_{dis}	Heat power of disturbances	0.43	kW
κ_P	Constant coefficient of the inverter AC's power	0.04	kW/Hz
κ_Q		0.12	kW/Hz
μ_P	Constant coefficient of the AC's refrigerating capacity	0.02	kW
μ_Q		−0.05	kW
Δf_W^{min}	The minimum dead band of the frequency deviation	0.01	Hz
Δf_W^{max}	The maximum dead band of the frequency deviation	0.03	Hz

5.5 Case Studies

5.5.1 Test System

The test model adopts the power system in Fig. 5.3, which includes the reheat steam generator, conventional loads and inverter ACs. The parameters of the temperature and the thermal model are based on the test data and the national standards in Hangzhou, China, on August 1st, 2015 [5]. The ambient temperature is 33 °C at 12:00 AM. The number of inverter ACs and corresponding rooms is 30,000. The living areas of these rooms are assumed to follow the normal distribution, where the mean value is 100 m^2 and the standard deviation is 40 m^2. The height of all the rooms is 2.5 m. The set temperatures of inverter ACs are distributed randomly between 22 and 26 °C to simulate different requirements of room temperature for various customers. The maximum allowable deviation of the room temperature ΔT_A^{max} is 1°C. Moreover, the frequency range of each inverter AC is 1~150 Hz. Time constant and the rate limiter of the compressor are 0.02 s and 10 Hz/s, respectively. The proportional gain θ and the integral gain η of the temperature controller are 0.52 Hz/°C and 0.032 Hz/(°C Δ s), respectively. The other controller, which connects the system frequency and the inverter AC's frequency, are set by the proportional parameter 200 and the integral parameter 0.02(1/s), respectively. Moreover, other constant parameters of the thermal model and electrical model are shown in Table 5.2.

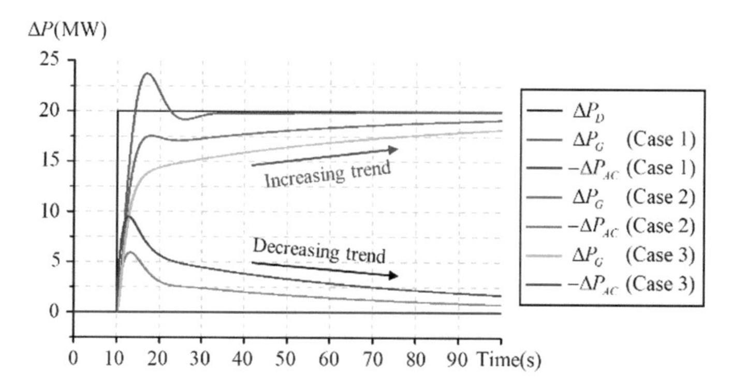

Fig. 5.10 Regulation power of the generator and inverter ACs in the under frequency scenario

The generation capacity of the reheat steam generator is 800 MW [9]. The generator inertia H is 10. The load-damping factor K_D is 1. The speed governor time constant T_g, reheat time constant T_r and turbine time constant T_t are 0.2 s, 7 s and 0.3 s, respectively. The power fraction of the high pressure turbine section F_{HP} is 0.3. The speed droop parameter R and the integral gain K are 0.05 and 0.50, respectively [9].

It is assumed that the system operates in the normal state and maintains the rated frequency at 50 Hz. The initial load is around 560 MW and the required capacity for FRS is 80 MW. The maximum allowable share of FRS provided by inverter ACs is 50%, which is 40 MW. Three cases are simulated: all the required capacity of FRS (80 MW) is provided by the generator in Case 1, while 20 and 40 MW of the required capacity of FRS are provided by inverter ACs in Case 2 and Case 3, respectively.

The frequency deviation is assumed to occur at 12:00 AM, when 20 MW loads are abruptly added to the system in the under frequency scenario and cut down in the over frequency scenario, respectively.

5.5.2 Simulation Results

The simulation results of the power deviations are shown in Figs. 5.10 and 5.11, respectively.

With the increasing share of inverter ACs in the FRS, the regulation power provided by inverter ACs becomes larger, while the corresponding regulation power provided by the generator becomes smaller. It illustrates that part of the regulation power can be supplied from inverter ACs in place of the generator, and thus, it shows that inverter ACs can be equivalent to the generator to provide FRS.

However, the regulation power provided by inverter ACs cannot last for a long time as the generator. As shown in Fig. 5.10, the regulation power of inverter ACs has a decreasing trend after reaching the required maximum capacity, which indicates that the operating power of inverter ACs rises again and ACs return to the normal

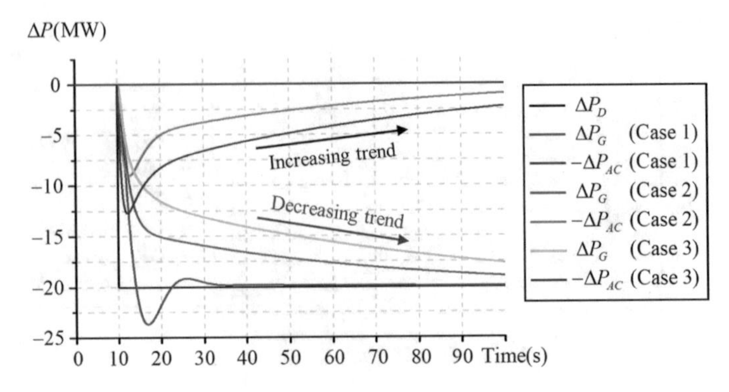

Fig. 5.11 Regulation power of the generator and inverter ACs in the over frequency scenario

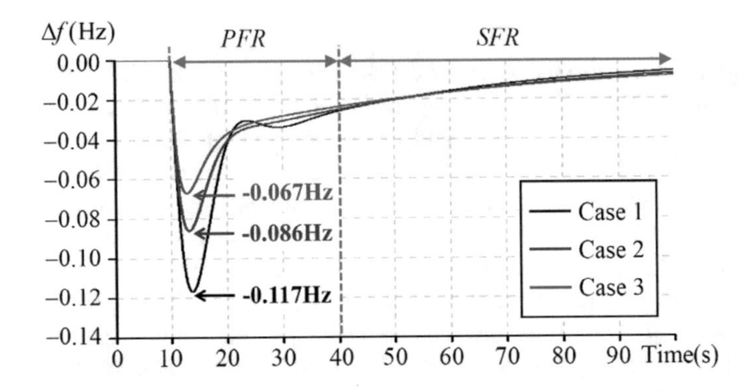

Fig. 5.12 Frequency deviation process in the under frequency scenario

operation state along with the recovery of the system frequency. Around 90 s later, inverter ACs' regulation power will be close to zero. By contrast, the regulation power provided by the generator has an increasing trend and will finally compensate for the regulation capacity provided by inverter ACs.

The fluctuation processes of the frequency deviation are shown in Figs. 5.12 and 5.13, respectively.

In the under frequency scenario, the maximum frequency deviation decreases from 0.117 Hz in Case 1 to 0.067 Hz in Case 3, when half of the frequency regulation capacity is provided by inverter ACs. In order to explain this observation, some evaluation parameters are shown in Tables 5.3 and 5.4.

It can be seen from the Tables 5.3 and 5.4 that the total regulation powers ($|\Delta P_G^{max}|+|\Delta P_{AC}^{max}|$) in the three cases are almost the same. The activation time of inverter ACs (AT_{AC}) is shorter than that of the generator (AT_G), where the AT_{AC} and AT_G are 3.19 s and 7.95 s in Case 2, respectively. It shows that inverter ACs can provide FRS more rapidly than the generator. The generator regulates its power generation through a series of processes, such as the speed governor process and the reheat steam turbine process, leading to a larger inertia compared with inverter ACs.

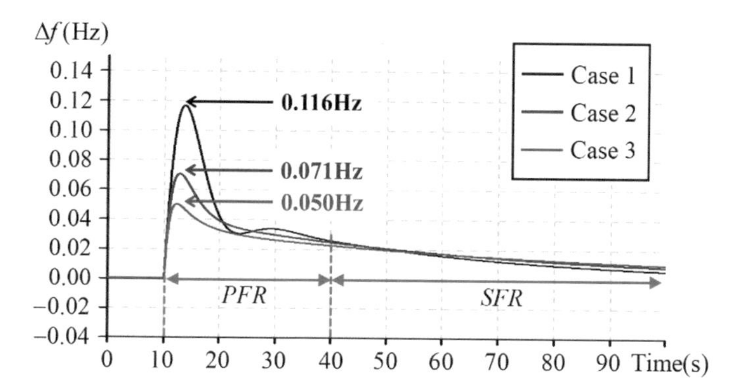

Fig. 5.13 Frequency deviation process in the over frequency scenario

Table 5.3 Simulation results in the under frequency scenario

Cases	ΔP_G^{max} (MW)	AT_G (s)	ΔP_{AC}^{max} (MW)	AT_{AC} (s)	Δf_{max} (Hz)	RT (s)
Case 1	23.704	6.98	0	N/A	−0.117	24.87
Case 2	17.545	7.95	−5.956	3.19	−0.086	21.19
Case 3	13.116	8.32	−9.491	2.81	−0.067	16.90

Table 5.4 Simulation results in the over frequency scenario

Cases	ΔP_G^{max} (MW)	AT_G (s)	ΔP_{AC}^{max} (MW)	AT_{AC} (s)	Δf_{max} (Hz)	RT (s)
Case 1	−23.703	6.97	0	N/A	0.116	24.80
Case 2	−15.045	7.68	9.006	2.80	0.071	19.35
Case 3	−11.955	8.75	12.745	2.27	0.050	12.94

Faced with the sudden power disturbance in power systems, the fast regulation speed is important and contributes to decreasing the frequency deviation.

The recovery time (RT) is defined as the time interval from the occurrence moment of the frequency deviation to the moment when the deviation is less than 0.06% of the rated frequency [33], which is mainly related to the regulation speed of the generator's power generation and the regulation speed of inverter ACs' power consumption. The RT s in the three cases are shorter than 30 s, which can meet the requirements of the recovery time in the practical power systems [31, 32].

Figure 5.14 shows the fluctuations of the room temperature. As for the inverter ACs whose set temperatures are 26 °C, the maximum fluctuation of the corresponding room temperature is less than 0.25 °C, as shown in Fig. 5.14 (Case 1, Case 2 and Case 3). Around 30 min later, the room temperature returns to the original set point, i.e., 26 °C. In the case that a certain room's temperature reaches the upper temperature limit during the process of providing FRS, the corresponding inverter AC will stop

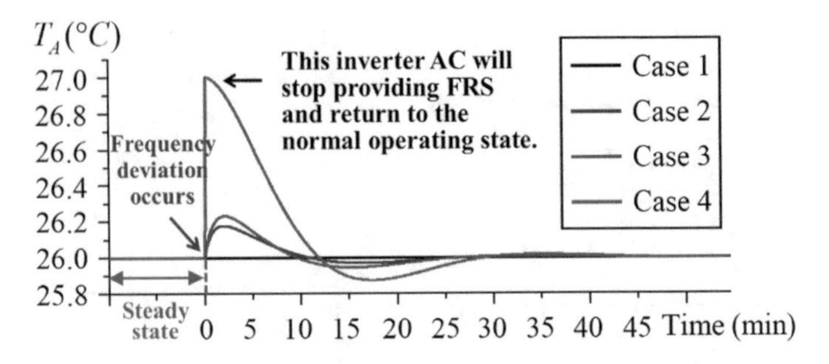

Fig. 5.14 Fluctuations of the room temperature in the under frequency scenario

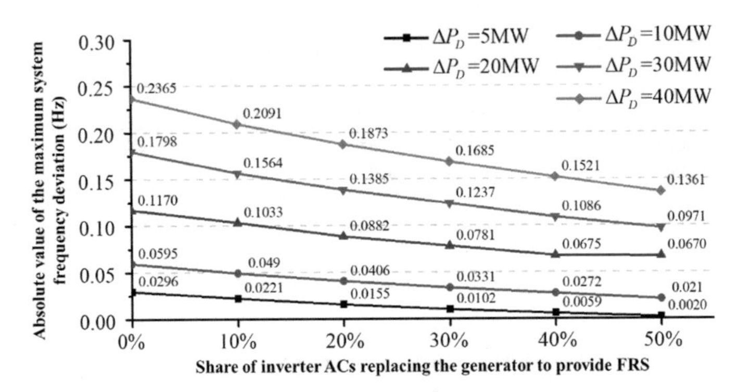

Fig. 5.15 The absolute value of the maximum system frequency deviation in the under frequency scenario

providing FRS and return to the normal operating state, as shown in Fig. 5.14 (Case 4). However, there would be few inverter ACs whose room temperatures reach the limit during the process of providing FRS, because the regulation period of FRS is short and thus, the operation of inverter ACs is only interrupted in a short time. Therefore, the service quality of the individual participants can be maintained.

More cases are simulated to verify the effectiveness of inverter ACs providing FRS. The initial parameters remain the same as the above three cases. The variable is the share of FRS provided by inverter ACs in the total required capacity (80 MW). Moreover, different deviations of loads (5, 10, 20, 30 and 40 MW) are considered, respectively.

The simulation results of the maximum frequency deviation are shown in Fig. 5.15. With the increasing of the sudden added loads, the frequency deviation becomes larger. Besides, inverter ACs can reduce the system frequency deviation by providing FRS. For example, the maximum frequency deviation decreases from 0.2365 to 0.1361 Hz under the same power deviation 40 MW.

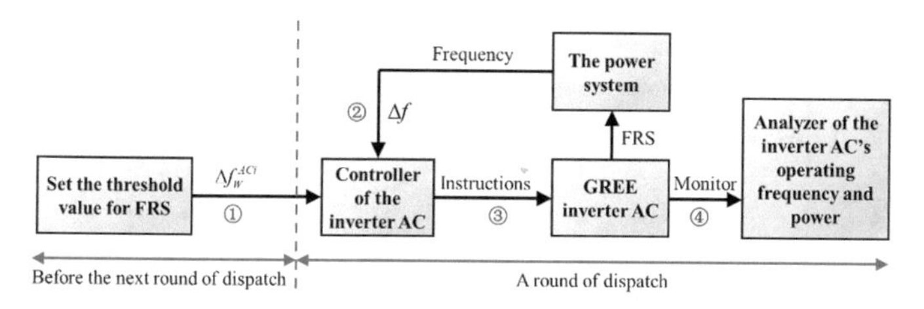

Fig. 5.16 Configuration of the experimental system

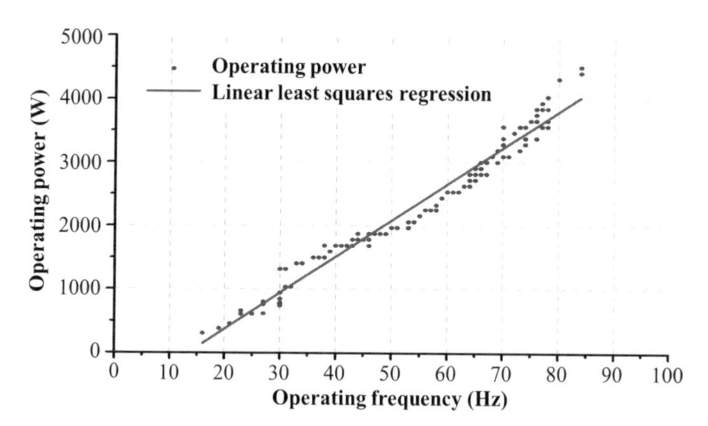

Fig. 5.17 The linear least squares regression of the inverter AC's operating power and frequency

5.5.3 *Experimental Results*

The experiment was developed on an inverter AC (GREE KFR-72LW/(72555) FNhAd-A3). The power supply voltage and rated frequency are 220 V and 50 Hz, respectively. The ambient temperature is 5 °C. The inverter AC operates in the heating mode and the set temperature is 26 °C.

As shown in Fig. 5.16, the threshold value Δf_W^{ACi} of the inverter AC to provide FRS is set on the controller before the next round of dispatch. When the new round of dispatch begins, the controller will keep monitoring the power system's frequency. If the system's frequency deviation Δf is larger than the threshold value Δf_W^{ACi}, the controller will send instructions to the inverter AC to adjust the operating frequency of the compressor. The operating frequency and power of the inverter AC can be monitored by the analyzer, whose sampling time interval is 0.5 s.

Figure 5.17 shows the relationship between the operating power and frequency of the inverter AC. It can be seen that the inverter AC's power is well fit with the operating frequency by the linear least squares approximation, where the slope is 57 W/Hz. Therefore, the inverter AC's operating power will rise/drop with the increase/decrease of its compressor's frequency. This observation verifies that the operating power of the inverter AC can be changed by adjusting the operating frequency of the compressor.

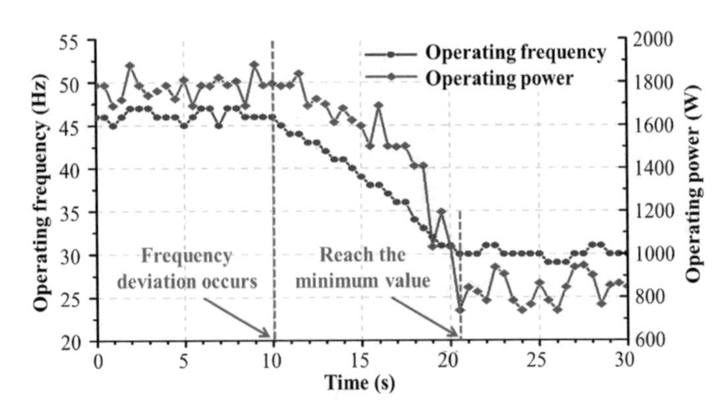

Fig. 5.18 The response process of the inverter AC for providing FRS

The threshold value Δf_W^{ACi} of the inverter AC is set to 0.01 Hz. The system's frequency deviation Δf is 0.05 Hz. The controller will send instructions to the inverter AC to adjust operating frequency of the compressor. As shown in Fig. 5.18, the compressor's operating frequency decreases from around 46 to 30 Hz, and the inverter AC's operating power also drops from around 1800 to 750 W. As shown in this experiment, the activation time (AT_{AC}) of the inverter AC as defined in this chapter is 11 s, which can meet the requirements of the response time for FRS.

5.6 Conclusions

The chapter presents the aggregation of inverter ACs as a traditional generator to provide FRS for the system. A thermal model of a room and an electrical model of an inverter AC considering the participation of FRS are developed. Based on this model, the inverter AC is equivalent to a reheat steam generator, including equivalent transfer functions, control parameters and evaluation criteria. In this manner, inverter ACs can be compatible with the existing control system and controlled just as traditional generators to provide FRS. A stochastic allocation method of the regulation sequence among inverter ACs is proposed to reduce the effect of FRS on customers. A hybrid control strategy, comprising the dead band control and hysteresis control, is also designed to reduce the frequency fluctuations of power systems. The simulation and practical results verified that the aggregation of inverter ACs can be equivalent to a generator to participate in FRS, while ensuring the requirement of customers' comfort. Besides, inverter ACs can be regulated more quickly, which makes up the generator's shortcoming on the regulation speed. The system stability gets enhanced when a certain share of FRS is provided by inverter ACs.

References

1. T. Strasser, F. Andrén, J. Kathan, C. Cecati, C. Buccella, P. Siano, P. Leitao, G. Zhabelova, V. Vyatkin, P. Vrba, V. Mařík, A review of architectures and concepts for intelligence in future electric energy systems. IEEE Trans. Ind. Electron. **62**(4), 2424–2438 (2015)
2. Z. Li, X. Wu, K. Zhuang, L. Wang, Y. Miao, B. Li, Analysis and reflection on frequncy characteristics of East China Grid after bipolar locking of 9.19 Jinping-Sunan DC transimission line. Autom. Electr. Power Syst. **41**(7), 149–155 (2017)
3. Administrative investigation report on the power failure 815, Executive Yuan, Taiwan, Republic of China, Technical Report (2017), http://www.ey.gov.tw
4. Y.G. Rebours, D.S. Kirschen, M. Trotignon, S. Rossignol, A survey of frequency and voltage control ancillary services—Part I: technical features. IEEE Trans. Power Syst. **22**(1), 350–357 (2007)
5. H. Hui, Y. Ding, W. Liu, Y. Lin, Y. Song, Operating reserve evaluation of aggregated air conditioners. Appl. Energy **196**, 218–228 (2017)
6. P. Siano, Demand response and smart grids—a survey. Renew. Sustain. Energy Rev. **30**, 461–478 (2014)
7. Y. Wang, N. Zhang, C. Kang, D.S. Kirschen, J. Yang, Q. Xia, Standardized matrix modeling of multiple energy systems. IEEE Trans. Smart Grid (2017), https://doi.org/10.1109/tsg.2017.2737662. (in press)
8. P. Palensky, D. Dietmar, Demand side management: Demand response, intelligent energy systems, and smart loads. IEEE Trans. Ind. Inform. **7**(3), 381–388 (2011)
9. J. Nanda, S. Mishra, L.C. Saikia, Maiden application of bacterial foraging-based optimization technique in multiarea automatic generation control. IEEE Trans. Power Syst. **24**(2), 602–609 (2009)
10. P. Siano, D. Sarno, Assessing the benefits of residential demand response in a real time distribution energy market. Appl. Energy **161**(7), 533–551 (2016)
11. J. Wang, H. Zhong, C. Tan, X. Chen, R. Rajagopal, Q. Xia, C. Kang, Economic benefits of integrating solar-powered heat pumps into a CHP system. IEEE Trans. Sust. Energy (2018), https://doi.org/10.1109/tste.2018.2810137. (in press)
12. H. Liu, Z. Hu, Y. Song, J. Wang, X. Xie, Vehicle-to-grid control for supplementary frequency regulation considering charging demands. IEEE Trans. Power Syst. **30**(6), 3110–3119 (2015)
13. A. Molina-Garcia, F. Bouffard, D.S. Kirschen, Decentralized demand-side contribution to primary frequency control. IEEE Trans. Power Syst. **26**(1), 411–419 (2010)
14. G. Benysek, J. Bojarski, R. Smolenski, M. Jarnut, S. Werminski, Application of stochastic decentralized active demand response (DADR) system for load frequency control. IEEE Trans. Smart Grid **99**, 1–8 (2016)
15. Y. Bao, Y. Li, Y. Hong, B. Wang, Design of a hybrid hierarchical demand response control scheme for the frequency control. IET Gener. Transm. Distrib. **9**(15), 2303–2310 (2015)
16. S. Weckx, R. D'Hulst, J. Driesen, Primary and secondary frequency support by a multi-agent demand control system. IEEE Trans. Power Syst. **30**(3), 1394–1404 (2014)
17. M. Isaac, D.P.V. Vuuren, Modeling global residential sector energy demand for heating and air conditioning in the context of climate change. Energy Policy **37**(2), 507–521 (2009)
18. Air conditioning consumes one third of peak electric consumption in the summer, Science Daily, Technical Report (2012), https://www.sciencedaily.com
19. AC makers betting on consumers' shift to inverter models, BusinessLine, Technical Report (2017), http://www.thehindubusinessline.com
20. Analysis on inverter air conditioners in China in Oct. 2015, Information network of Chinese business, Technical Report (2015), http://www.askci.com
21. What is Inverter Technology AC, Bijli Bachao, Technical Report (2017), https://www.bijlibachao.com
22. M. Song, C. Gao, H. Yan, J. Yang, Thermal battery modeling of inverter air conditioning for demand response. IEEE Trans. Smart Grid **99**, 1–13 (2017)

23. S. Shao, W. Shi, X. Li, H. Chen, Performance representation of variable-speed compressor for inverter air conditioners based on experimental data. Int. J. Refrig. **27**(8), 805–815 (2004)
24. W. Zhang, J. Lian, C.Y. Chang, K. Kalsi, Aggregated modeling and control of air conditioning loads for demand response. IEEE Trans. Power Syst. **28**(4), 4655–4664 (2013)
25. N. Mahdavi, J. H. Braslavsky, C. Perfumo, Mapping the effect of ambient temperature on the power demand of populations of air conditioners. IEEE Trans. Smart Grid **99**, 1–10 (2016)
26. N. Lu, An evaluation of the HVAC load potential for providing load balancing service. IEEE Trans. Smart Grid **3**(3), 1263–1270 (2012)
27. J. Grainger, W.D. Stevenson, *Power System Analysis*, 1st edn. (McGraw-Hill, Michigan, U.S.A, 1994)
28. Z. Han, *Power System Analysis*, 5th edn. (Zhejiang University Press, Hangzhou, China, 2013)
29. H. Hui, Y. Ding, M. Zheng, Equivalent modeling of inverter air conditioners for providing frequency regulation service. IEEE Trans. Ind. Electron. **66**(2):1413-1423 (2019).
30. J. Wang, C. Zhang, Y. Jing, D. An, Study of neural network PID control in variable-frequency air-conditioning system, in IEEE International Conference on Control and Automation, Guangzhou, China, 30 May–1 June, 2007, pp. 317–322
31. Continental Europe Operation Handbook, Policy 1-Load Frequency Control and Performance, entsoe, Technical Report (2018), https://www.entsoe.eu/fileadmin/user_upload/_library/publications/entsoe/Operation_Handbook/Policy_1_final.pdf
32. Continental Europe Operation Handbook, Appendix 1-Load Frequency Control and Performance, entsoe, Technical Report (2018), https://www.entsoe.eu/fileadmin/user_upload/_library/publications/entsoe/Operation_Handbook/Policy_1_Appendix%20_final.pdf
33. Frequency Response Requirements-Phase 1 (ER16-1483), California ISO, 21 April, 2016, https://www.caiso.com/Documents/Apr21_2016_FrequencyResponseRequirements_Phase1_ER16-1483.pdf

Chapter 6
Integration of Flexible Heating Demand into the Integrated Energy System

6.1 Introduction

The previous chapters have well analyzed the demand response of the ACs and proposed a framework for aggregating the demand response potential to provide the flexibility required by the power system. Such demand response potential is not restricted to the air-conditioners. The other energy conversion and energy storage devices can also be utilized to provide the demand response. Moreover, the demand response can be used not only in the power system but also in the integrated energy system which involves different kinds of energy forms, such as electricity, heat and so on.

This chapter expandes the demand response to the heat and power integrated energy system (HE-IES). HE-IES, based on combined heat and power (CHP), is one of the most important forms of IES [1]. Countries around the world have made great efforts to develop CHP. In Denmark, CHP covers about 40% of the demand for space heating [2]. In northern China, CHP based district heating (DH) has been installed in more than 300 cities, serving 40% of China's population [3]. However, the inharmony between the high-penetration wind power and the wide use of CHP has become a challenge for operating the energy systems [4]. On the one hand, the fluctuation of wind power makes electricity price vary significantly during the operating day. Thus, the profit of CHP units can be dramatically reduced in periods of low electricity price resulting from large wind power production [5]. On the other hand, the electricity generation of CHP units is constrained by their heat production, which must target on customers' heat demand. This leads to high wind power curtailment when the electricity production of CHP units covers most of the electricity demand during the off-peak hours [3].

Facing the challenges of integrating wind power with CHP units, a lot of researches have been conducted on increasing the flexibility of CHP based energy system. These works are mostly focused on decoupling the heat production from electricity generation by coupling CHP units with thermal storage or electric heating systems (EHS), such as heat pumps and electric boilers [6, 7]. However, these measures are

Y. Ding et al., *Integration of Air Conditioning and Heating into Modern Power Systems*,
https://doi.org/10.1007/978-981-13-6420-4_6

more appropriate for relatively small CHP units used for district heating. For the centralized and high-capacity CHP units, it would be difficult to install the thermal stores of corresponding capacity. Moreover, the utilization of the heat storage capacity of the district heating network for increasing the flexibility has been proposed in [8]. Besides the measures from the production and network sides, the demand-side resources hold untapped potential for increasing the operational flexibility of CHP based IES. Given that the operation of CHP units is usually constrained by their heat output, encouraging customers to adjust their energy demand, especially the heat demand, in response to the supply conditions is crucial for increasing the flexibility of the energy system. Moreover, it has become a realistic possibility since more and more heating systems are transformed from constant flow systems to variable flow systems [9].

Indeed, the integration of multiple energy systems imparts flexibility to customers' energy demand. In the integrated energy system, customers have multiple options to fulfill their energy demand [10, 11]. For instance, their heating demand can be supplied through the heat power from the heat networks or the electric heating devices. Since alternate cost pricing is a common practice pricing mechanism used in DH system and electric heat pumps have come to be a price competitive alternative [12], it would be economically feasible for customers to switch to electric heating during low-electricity-price periods. Such built-in flexibility enables customers to adjust their energy demand in response to the supply conditions to reduce their energy bills. In this way, the independent system operator (ISO) will also obtain more balancing resources for maintaining the energy balance and improving the operation flexibility of the energy system. For example, when the electricity is oversupplied due to the increased electricity production from the wind power, the electricity price can be relatively low and customers therefore may increase the use of electric heating. Then, the aggregated demand for electricity increases while the demand for heat decreases. CHP units therefore can decrease the amount of heat and electricity produced. In this way, the rebalance is achieved without the curtailment of wind power and customers also benefit financially.

The potential for flexible multiple energy systems to provide demand response (DR) have been illustrated in [13–15]. These studies are more focused on identifying and quantifying their electricity shifting potential to participate in real-time DR programs. Moreover, the utilization of the demand response for releasing the heat production constraints in the HE-IES has not been discussed yet. This chapter investigates the utilization of customers' flexible energy demand, including both heat demand and electricity demand, to provide additional balancing resources for maintaining the energy supply and demand balance and avoiding the wind power curtailment. Customer aggregators are introduced to purchase energy from the centralized energy systems for supplying customer's energy demand in the most cost-effective way. Meanwhile, by controlling customers' energy consumption behaviors, aggregators can adjust their energy demand in response to supply conditions.

It is assumed that both electricity energy system and heat energy system are managed by a single ISO and all the aggregators seek to minimize their energy costs. Incorporating the aggregators' flexible energy demand into the central energy dispatch model therefore forms a two-level optimization problem (TLOP), where the ISO maximizes social welfare subject to aggregators' strategies in which aggregators adjust their energy demand to minimize the energy purchase cost. Moreover, the low-level problems are linearized based on several reasonable assumptions. KKT conditions of the low-level problems are then transformed into energy demand as explicit and piecewise-linear functions of electricity prices corresponding to the demand bid curves. In this way, it requires each aggregator to submit only a demand bid to run the centralized energy dispatch. All other parameters pertaining to the energy consumption models are internalized in the bid curves. After that, the TLOP problem is transformed into an extended optimal power flow (OPF) problem, in which electricity energy and heat energy are jointly optimized.

Simulation results find that the customers' flexible energy demand can serve as a tool to further bridge the heat and electricity energy systems from the consumption side which enables the co-optimization of the two kinds of energy systems. The illustrative results demonstrate that not only the volatility of electricity price and wind power curtailment can be reduced, but also the social welfare can be increased remarkably. This chapter includes research related to the modeling and integration of flexible demand in heat and electricity integrated energy system by [16].

6.2 Heat and Electricity Integrated Energy System

6.2.1 Description of the HE-IES

The proposed HE-IES is a combination of CHP based heating system and electric power system, which supplies heat and electricity for customers. The schematic graph of the HE-IES is shown in Fig. 6.1.

The system involves heat and electricity generating units, heat and electricity grids, and customers' energy demand as shown in Fig. 6.1. Generally, the generating units in the HE-IES can be divided into electricity-only units, cogenerating units and heat-only units. The electricity-only units include conventional electricity generating units, such as thermal power generating units (TPP), and renewable energy power generating units (RES power, such as wind power). The cogeneration units are CHP units, and the heat-only units usually refer to boilers. Moreover, the heat and electricity production of a CHP unit are coupled, which must stay within the feasible operating area [6].

The heat produced by the CHP units is delivered to the customers through the DH networks, to which the CHP units and customers' buildings are connected. After generation, the heat is distributed to the customers via the DH network of pipes. At customer level the heat network is usually connected to the central heating system of

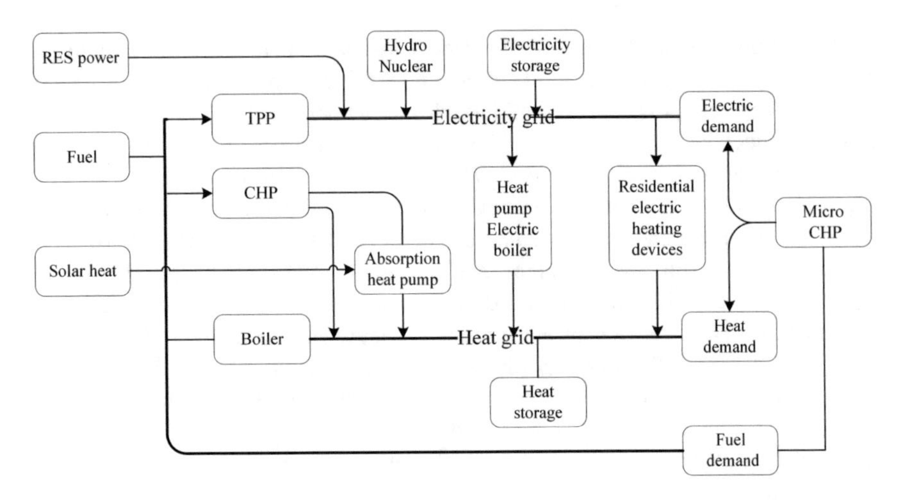

Fig. 6.1 Schematic graph of the HE-IES

the dwellings via substations [17]. In a DH system intended to supply a customers' energy requirement, two parameters can be controlled: supply temperature and flow rate [18]. Moreover, more and more DH networks are transformed to variable flow systems, which allows the heat output to change when some of the customers' heating demand changes. It is suggested that modern DH schemes give customers just as much control as individual gas boilers and could be very efficient [17].

6.2.2 Modelling the Customer Aggregators' Energy Demand

Customer aggregators as an independent entity is in charge of purchasing energy from the energy market and supplying the downstream demands. The aggregator can be modelled as a heat and electricity integrated energy consumption node, as shown in Fig. 6.2.

(1) **Energy Demand of the Aggregator**

Energy demand of the aggregator is defined as the demand for the imported electricity and heat from the energy grids. The electricity demand of the aggregator is equal to the sum of the electricity demand of all the customers, which is expressed as:

$$L_{e,i} = \sum\nolimits_{j \in \Xi_i} L_{e,j} \tag{6.1}$$

where Ξ_i identifies the set of customers served by the aggregator i.

The heat demand of the aggregator can be calculated as:

$$L_{h,i} = m_i C_w (T_s - T_r) \tag{6.2}$$

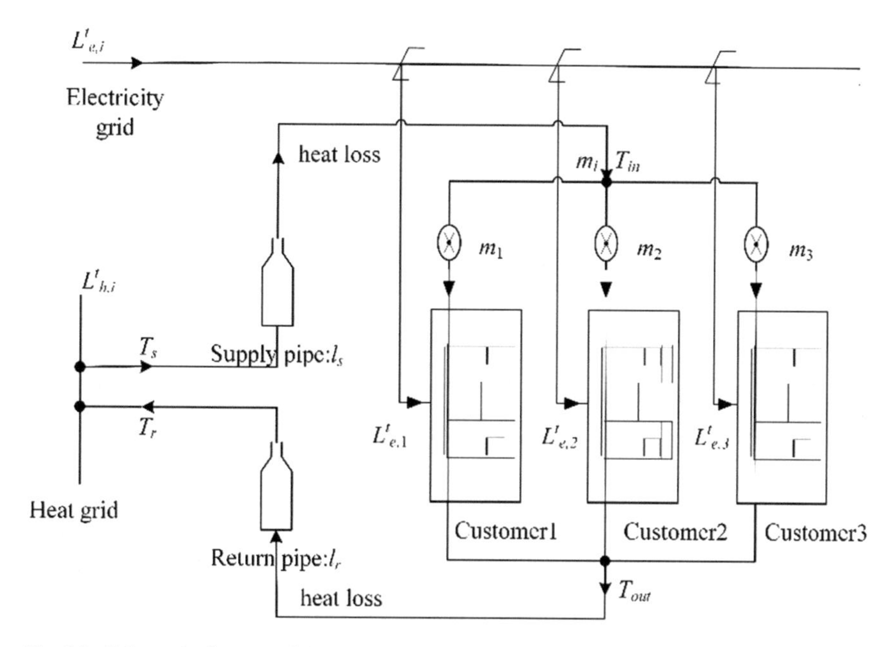

Fig. 6.2 Schematic diagram of the aggregator' energy demand

(2) Heating Network

A heat source supplies heat by means of hot water or steam, and then the heat power is delivered to the customers through the heating network. A heating network consists of supply pipes and return pipes. Heating systems can be controlled with constant flow and constant temperature, constant flow and variable temperature, variable flow and constant temperature, or variable flow and variable temperature [17]. Moreover, more district heating systems are transformed from constant flow systems to variable flow system [19]. In this chapter, it is assumed that the heating systems are operated with variable flow and constant temperature, i.e. Ts is maintained constant and m_i may vary within the selected range.

The temperature drops exponentially during water flow in pipes [20], hence:

$$\begin{aligned} T_{in} &= (T_s - T_a)e^{-hl_s/c_w m_i} + T_a \\ T_{out} &= (T_r - T_a)e^{hl_r/c_w m_i} + T_a \end{aligned} \tag{6.3}$$

where l_s and l_r are the length of the supply pipe and return pipe, respectively. Usually, it can be assumed that $l_s = l_r = l$.

Moreover, as we can see from Fig. 6.2, the heat loss in the network can be expressed as:

$$P_h^{loss} = C_w m_i (T_s - T_r) - C_w m_i (T_{in} - T_{out}) \tag{6.4}$$

As shown in (6.3) and (6.4), the heat loss is related to the gross water flow (which is dependent on the heating demand), ambient temperature, the length of the heat pipe, and so on.

The gross water flow of the heating system equals to the sum of the flow of the pumps at customer sites, which is expressed as:

$$m_i = \sum_{j \in \Xi_i} m_j \tag{6.5}$$

Moreover, it is defined that:$m_j = R_j m_j^{set}$, where m_j^{set} is the rated water flow, and R_j is the relative water flow ratio.

(3) **Electricity Power Balance**

The electricity demand of the customer j can be divided to heating-demand $L_{es,j}$ and non-heating demand $L_{e0,j}$. The non-heating demand includes the lighting, pumps, fans and other electric appliances. The electricity demand $L_{e,j}$ is the sum of the heating-demand and non-heating demand, which is expressed as:

$$L_{e,j} = L_{es,j} + L_{e0,j} \tag{6.6}$$

(4) **Heat Power Balance**

There is a direct relationship between heating load and the temperature difference between the inside and outside of the customers' building. The more the temperature difference is, the more the heating demand is. Generally, indoor temperature of customer j is centrally controlled and is assumed to be the same as that of other customers. The heating load of customer j is estimated as:

$$L_{h0,j} = H_j V_j (T_n - T_a) \tag{6.7}$$

The heat supplied to customer j comes from heating radiators and electric heating devices. The total value of heat injected into the customer's building $H_{in,j}$ can be calculated as the following:

$$H_{in,j} = H_{es,j} + H_{net,j} = \eta_j L_{es,j} + m_j C_w (T_{in} - T_{out}) \tag{6.8}$$

where $H_{es,j}$ and $H_{net,j}$ are the heat release rate of the electric heating device and the radiator installed for customer j, respectively. $L_{es,j}$ is the electric power input of the electric heating and η_j represents its efficiency.

Therefore, the response of the indoor temperature of customer j can be expressed as:

$$\rho_j C_w V_j \frac{dT_n}{dt} = \eta_j L_{es,j} + m_j C_w (T_{in} - T_{out}) - L_{h0,j} \tag{6.9}$$

6.3 TLOP-Formulation of the Dispatch Model

In this chapter, it is assumed that both the electricity system and heat system are managed by a single ISO. The ISO runs a centralized energy dispatch model for the dispatch of heat energy and electricity energy. As aforementioned, the integration of heat energy and electricity energy imparts flexibility to customer's energy demand. The flexibility can be further managed and aggregated by the aggregators. Moreover, the energy market can provide economic incentives to offer flexibility through real-time price mechanism. Aggregators in charge of controlling customer's energy consumption pattern could take advantage of these incentives for minimizing the energy purchase cost. The aggregators' response, in return, provides additional flexibility to power system operators, which is necessary to the transition towards more renewable generation.

However, it should be noted that the aggregators' adjusting their energy demand introduces additional uncertainty to the energy system operation. Usually, it can be assumed that all aggregators are rational market participants who try to minimize the energy purchase cost. As a result, the centralized dispatch model forms a two-level optimization problem where the ISO maximizes the social welfare based on all bids in the market subject to a sub solution where the aggregators minimize the energy purchase cost based on the energy prices. To be specific, aggregators are supposed to send information on their energy consumption models to the ISO in the proposed approach. The ISO runs the centralized energy dispatch model with the consideration of the aggregators' optimal strategy. In this way, the clearing results guarantee not only the optimal solution for the total energy system, but also the minimum costs for all aggregators. Hence, the convergence to an equilibrium among all aggregators can be guaranteed.

The formulation of the two-level optimization is shown in (6.10)–(6.28).

$$\begin{aligned} \underset{G^t_{e,gi},G^t_{e,gj},G^t_{h,gj}}{Maximize}\ F = & \sum_{t=1}^{NT} \Big(\sum_{i=1}^{Na} B_i(L^t_{e,i}, L^t_{h,i}) \\ & - \sum_{gi=1}^{Ngi} C^t_{gi}(G^t_{e,gi}) - \sum_{gj=1}^{Ngj} C^t_{gj}(G^t_{e,gj}, G^t_{h,gj})\Big) \end{aligned} \tag{6.10}$$

Subject to:

$$\begin{aligned} & \sum_{gi\in\Psi_n} G^t_{e,gi} + \sum_{gj\in\Psi_n} G^t_{e,gj} - \sum_{i\in\Psi_n} L^t_{e,i} \\ & = \sum_{p\in\Theta_n} V_n V_p \big[G_{np}\cos(\delta^t_n - \delta^t_p) + B_{np}\sin(\delta^t_n - \delta^t_p)\big] \end{aligned} \tag{6.11}$$

$$\sum_{n\in\Pi_d} \sum_{gj\in\Psi_n} G^t_{h,gj} = \sum_{n\in\Pi_d} \sum_{i\in\Psi_n} L^t_{h,i} \quad \forall t \tag{6.12}$$

$$\underline{G^t_{e,gi}} \le G^t_{e,gi} \le \overline{G^t_{e,gi}} \quad \forall t,\ \forall gi \tag{6.13}$$

$$\underline{G^t_{e,gj}} \le G^t_{e,gj} \le \overline{G^t_{e,gj}} \quad \forall t,\ \forall gj \tag{6.14}$$

$$\underline{G_{h,gj}^{t}} \le G_{h,gj}^{t} \le \overline{G_{h,gj}^{t}} \quad \forall t, \ \forall gj \tag{6.15}$$

$$G_{h,gj}^{t} = \Upsilon_{he,gj} G_{e,gj}^{t} \tag{6.16}$$

$$-T_{np}^{\max} \le T_{np} \le T_{np}^{\max}, \quad \forall p \in \Theta_n \tag{6.17}$$

$$-\pi \le \delta_n^t \le \pi \qquad \forall t, \ \forall n \tag{6.18}$$

$$(L_{e,i}^{t}, L_{h,i}^{t}) \in \arg\ minimize\ p_e^t L_{e,i}^{t} + p_h L_{h,i}^{t} \tag{6.19}$$

subject to:

$$L_{e,i}^{t} = \sum\nolimits_{j \in \Xi_i} L_{e,j}^{t} \tag{6.20}$$

$$m_i^t = \sum\nolimits_{j \in \Xi_i} m_j^t \tag{6.21}$$

$$m_j^t = R_j^t m_j^{set}, \quad \forall t, \forall j \tag{6.22}$$

$$\eta_j L_{es,j}^{t} + m_j^t C_w (T_{in}^{t} - T_{out}^{t}) - L_{h0,j}^{t} = 0 \ \forall t, \forall j \tag{6.23}$$

$$T_{in}^{t} = (T_s - T_a^t) e^{hl/c_w m_i^t} + T_a^t \tag{6.24}$$

$$T_{out}^{t} = (T_r - T_a^t) e^{hl/c_w m_i^t} + T_a^t \tag{6.25}$$

$$L_{e,j}^{t} = L_{es,j}^{t} + L_{e0,j}^{t} \quad \forall j \in \Xi_i \tag{6.26}$$

$$\underline{R} \le R_j^t \le \overline{R} \quad \forall t, \forall j \tag{6.27}$$

$$p_e^t = \lambda_n^t \tag{6.28}$$

The high-level problem (6.10)–(6.18) represents the centralized energy dispatch with the target of maximizing the social welfare. Equations (6.11) and (6.12) represent the power balance and heat balance, respectively. Since this chapter is focused on the steady state, the heating network in a heating district is equivalent to a single node, where only the heat balance is considered, as shown in (6.12) [21]. Equations (6.13)–(6.15) are power output bounds for the generating units. Constraints (6.17) enforces the transmission capacity limits of each line. Constraints (6.18) stands for angle bounds for each node. $n \in \Pi_d$ identifies the node belongs to heating district $d.gi \in \Psi_n$ identifies the thermal power generating unit located at bus n. $p \in \Theta_n$ identifies the bus p connected to bus n. $\Upsilon_{he,gj}$ is the heat-to-electricity ratio of the gj-th CHP unit. It should be noted that the heat demand and electricity demand $(L_{e,i}^{t}, L_{h,i}^{t})$ are endogenously generated within the lower-level problem.

The low-level problem (6.19)–(6.28) represents strategy of aggregator i for minimizing the energy purchase cost. Equation (6.23) enforces a constant indoor temperature of customers' building to ensure comfort. λ_n^t is the dual variable on the equilibrium constraint, obtained from the high-level problem.

The most common approach to solving the TLOP is to replace the sub-problems by their KKT conditions [22]. In this way, the TLOP can be written as a standard optimization problem. However, it requires the detailed parameters pertaining to aggregators' energy consumption model when ISO runs the centralized energy dispatch program, as can be seen from its formulation. Considering that the aggregators can be in large number and widely distributed, it would be impractical for aggregators to send detailed energy consumption models to the ISO. Moreover, considering all the parameters is bound to increase the computation burden, it is necessary to modify the TLOP by simplifying the KKT conditions of the low-level problems.

6.4 Simplifying the Sub-problems' KKT Conditions

As discussed above, a TLOP is presented in which aggregators try to minimize their energy purchase cost under the constraint that their dispatch and price are determined by the centralized energy dispatch. Usually, the TLOP can be solved by representing the sub-problems with the KKT conditions. This chapter tries to linearize the sub-problems by linearizing the nonlinear constraints. After that, the KKT conditions of the sub-problems can be represented by a set of linear constraints, which are modelled as the energy demands as explicit linear functions of electricity price corresponding to the demand bid curves. In this way, the TLOP is transformed to a standard optimization problem, which requires aggregators to only submit a demand bid to run the centralized energy dispatch program.

In this section, reasonable assumptions are made for the simplification of the low-level problems. More specifically, the inequality constrains are linearized, making it possible to find the linear relationships between the energy demand $L_{e,i}^t$, $L_{h,i}^t$ and electricity price in the KKT conditions.

As aforementioned, the temperature distribution along the heating pipes can be expressed as (6.3).

In practice, $\mu = hl_r / c_w m_i^t$ is very small. Using the equivalent infinitesimal $\lim_{\mu \to 0} e^{\mu} = 1 + \mu$, (6.3) can be approximately written as [20]:

$$\begin{aligned} T_{in}^t &= (T_s - T_a^t)(1 - hl / c_w m_i^t) + T_a^t \\ T_{out}^t &= (T_r - T_a^t)(1 + hl / c_w m_i^t) + T_a^t \end{aligned} \tag{6.29}$$

Moreover, it is reasonable to assume that the rated water flow is determined by and proportional to the customers' heating load, that is:

$$m_j^{set} \Big/ \sum\nolimits_{j \in \Xi_i} m_j^{set} = L_{h0,j}^t \Big/ \sum\nolimits_{j \in \Xi_i} L_{h0,j}^t = K_j \tag{6.30}$$

Following this assumption, the heat power balance constraint (24) can be expressed as

$$\eta_j L^t_{es,j} + C_w\left[(T_s - T_r)m^t_j - \frac{m^t_j}{m^t_i}(T_s + T_r - 2T^t_a)\frac{hl}{C_w}\right] = L_{h0,j} \tag{6.31}$$

Moreover, m^t_i is approximated as:

$$m^t_i = \sum\nolimits_{j\in\Xi_i} m^t_j = \sum\nolimits_{j\in\Xi_i} R^t_j m^t_j \approx R_{av}\sum\nolimits_{j\in\Xi_i} m^{set}_j \tag{6.32}$$

where $R_{av} = (\underline{R} + \overline{R})/2$.

Hence, (6.23)–(6.25) are replaced by:

$$\eta_j L^t_{es,j} + C_w\left[(T_s - T_r)m^{set}_j - \frac{K_j}{R_{av}}(T_s + T_r - 2T^t_a)\frac{hl}{C_w}\right]R^t_j = L_{h0,j} \tag{6.33}$$

Since the nonlinear constraints in the low-level problem is linearized, the low-level problem is transformed to a linear optimization problem. The KKT conditions of the low-level problems therefore can be transformed to heat demand and electricity demand as explicit piecewise-linear functions of electricity price.

To facilitate the understanding of the simplified optimal conditions, a case is illustrated where $j = \{1, 2, 3\}$. The optimal strategy of the aggregator can be described as:

$$\begin{aligned}
&\underset{L^t_{e,i},L^t_{h,i},L^t_{es,j},R^t_j}{minimize} \quad p^t_e L^t_{e,i} + p_h L^t_{h,i} \\
&s.t. \quad L^t_{e,i} = \sum\nolimits_{j=1,2,3}(L^t_{es,j} + L^t_{e0,j}) \\
&\qquad L^t_{h,i} = \sum\nolimits_{j=1,2,3} R^t_j m^{set}_j C_w(T_s - T_r) \\
&\qquad \eta_j L^t_{es,j} + \zeta^t_j R^t_j - L^t_{h0,j} = 0 \quad \forall t, \forall j = 1,2,3 \\
&\qquad \underline{R} \le R^t_j \le \overline{R} \qquad \forall t, \forall j = 1,2,3
\end{aligned} \tag{6.34}$$

where $\zeta^t_j\left[(T_s - T_r)C_w m^{set}_j - \frac{K_j}{R_{av}}(T_s + T_r - 2T^t_a)hl\right]$.

The Lagrange function of the optimization problem is expressed as:

$$\begin{aligned}
L &= p^t_e L^t_{e,i} + p_h L^t_{h,i} + \lambda_e(L^t_{e,i} - \sum\nolimits_{j=1,2,3}(L^t_{es,j} + L^t_{e0,j})) \\
&+ \lambda_h(L^t_{h,i} - \sum\nolimits_{j\in\Xi_i} R^t_j m^{set}_j C_w(T_s - T_r)) \\
&+ \sum\nolimits_{j=1,2,3}(\lambda_{eh,j}(\eta_j P^t_{es,j} + \zeta^t_j R^t_j - L^t_{h0,j})) \\
&+ \sum\nolimits_{j=1,2,3}(\underline{\mu_{R,j}}(-R^t_j + \underline{R^t_j})) + \sum\nolimits_{j=1,2,3}(\overline{\mu_{R,j}}(R^t_j - \overline{R^t_j}))
\end{aligned} \tag{6.35}$$

The KKT conditions then can be expressed as [23]:

$$\partial L / \partial L_{e,i}^{t} = p_e^t + \lambda_e = 0$$
$$\partial L / \partial L_{h,i}^{t} = p_h + \lambda_h = 0$$

for $j = 1, 2, 3$:

$$\begin{cases} \partial L / \partial L_{es,j}^{t} = -\lambda_e + \lambda_{eh,j}\eta_j = 0 \\ \partial L / \partial R_j^t = -m_j^{set} C_w (T_s - T_r)\lambda_h + \zeta_j^t \lambda_{eh,j} - \underline{\mu_{R,j}} + \overline{\mu_{R,j}} = 0 \end{cases} \tag{6.36}$$

Define $\gamma_{eh,j} = \eta_j (T_s - T_r) C_w m_j^{set} \big/ \zeta_j^t$, and assume $\gamma_{eh,1} > \gamma_{eh,2} > \gamma_{eh,3}$. The optimal electricity demand and heat demand of the aggregator derived from the KKT conditions are expressed in (6.37) and (6.38), respectively.

$$L_{e,i}^t = \begin{cases} L_{e0,i}^t + \underline{L_{es,1}^t} + \underline{L_{es,2}^t} + \underline{L_{es,3}^t} & : p_e^t > \gamma_{eh,1} p_h \\ L_{e0,i}^t + L_{es,1}^t(R_1^t) + \underline{L_{es,2}^t} + \underline{L_{es,3}^t} & : \underline{R} < R_1^t < \overline{R}\ ,\ p_e^t = \gamma_{eh,1} p_h \\ L_{e0,i}^t + \overline{L_{es,1}^t} + \underline{L_{es,2}^t} + \underline{L_{es,3}^t} & : \gamma_{eh,1} p_h > p_e^t > \gamma_{eh,2} p_h \\ L_{e0,i}^t + \overline{L_{es,1}^t} + L_{es,2}^t(R_2^t) + \underline{L_{es,3}^t} & : \underline{R} < R_2^t < \overline{R}\ ,\ p_e^t = \gamma_{eh,2} p_h \\ L_{e0,i}^t + \overline{L_{es,1}^t} + \overline{L_{es,2}^t} + \underline{L_{es,3}^t} & : \gamma_{eh,2} p_h > p_e^t > \gamma_{eh,3} p_h \\ L_{e0,i}^t + \overline{L_{es,1}^t} + \overline{L_{es,2}^t} + L_{es,3}^t(R_3^t) & : \underline{R} < R_3^t < \overline{R}\ ,\ p_e^t = \gamma_{eh,3} p_h \\ L_{e0,i}^t + \overline{L_{es,1}^t} + \overline{L_{es,2}^t} + \overline{L_{es,3}^t} & : p_e^t < \gamma_{eh,3} p_h \end{cases} \tag{6.37}$$

where $L_{e0,i}^t = L_{e0,1}^t + L_{e0,2}^t + L_{e0,3}^t$, $L_{es,j}^t(R_j^t) = (L_{h0,j}^t - \zeta_j^t R_j^t)/\eta_j$, $\underline{L_{es,j}^t} = (L_{h,j}^t - \zeta_j^t \overline{R})/\eta_j$, $\overline{L_{es,j}^t} = (L_{h,j}^t - \zeta_j^t \underline{R})/\eta_j$.

$$L_{h,i}^t = \begin{cases} \zeta_1^t \overline{R} + \zeta_2^t \overline{R} + \zeta_3^t \overline{R} \quad p_e^t > \gamma_{eh,1} p_h \\ \zeta_1^t R_1^t + \zeta_2^t \overline{R} + \zeta_3^t \overline{R}\ :\ \underline{R} < R_1^t < \overline{R}\ p_e^t = \gamma_{eh,1} p_h \\ \zeta_1^t \underline{R} + \zeta_2^t \overline{R} + \zeta_3^t \overline{R}\ :\ \gamma_{eh,1} p_h > p_e^t > \gamma_{eh,2} p_h \\ \zeta_1^t \underline{R} + \zeta_2^t R_2^t + \zeta_3^t \overline{R}\ :\ \underline{R} < R_2^t < \overline{R}\ p_e^t = \gamma_{eh,2} p_h \\ \zeta_1^t \underline{R} + \zeta_2^t \underline{R} + \zeta_3^t \overline{R}\ :\ \gamma_{eh,2} p_h > p_e^t > \gamma_{eh,3} p_h \\ \zeta_1^t \underline{R} + \zeta_2^t \underline{R} + \zeta_3^t R_3^t\ :\ \underline{R} < R_3^t < \overline{R}\ p_e^t = \gamma_{eh,3} p_h \\ \zeta_1^t \underline{R} + \zeta_2^t \underline{R} + \zeta_3^t \underline{R}\ :\ p_e^t < \gamma_{eh,3} p_h \end{cases} \tag{6.38}$$

The electricity demand described in graphical mode is shown in Fig. 6.3 where the abscissa stands for the electricity demand $L_{e,i}^t$ and the ordinate stands for electricity price p_e^t.

As shown in Fig. 6.3, the electricity demand is expressed in a decreasing piecewise-linear function of the electricity price corresponding to a demand bid curve [24].

Heat demand can be expressed in a manner analogous to the electricity demand as shown in Fig. 6.4. A heat demand function is described in Fig. 6.4, which is mathematically similar to the electricity demand, except the demand function increases as the electricity price increases.

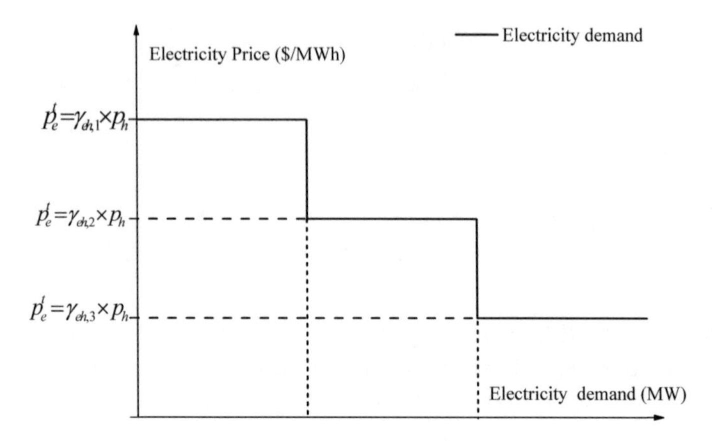

Fig. 6.3 Electricity demand derived from the KKT conditions

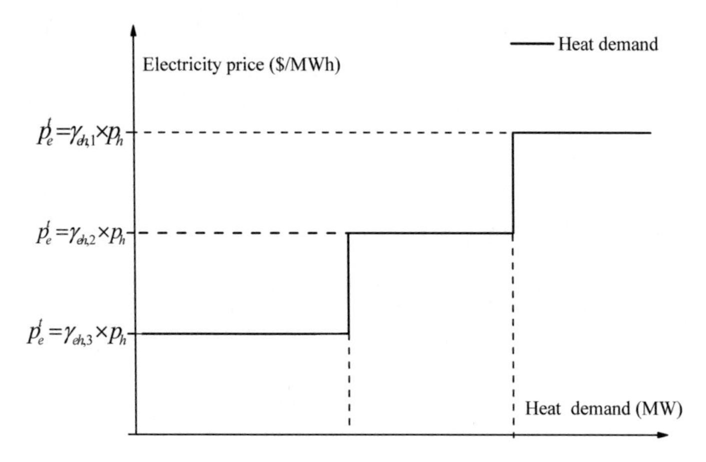

Fig. 6.4 Heat demand derived from the KKT conditions

In this way, KKT conditions are represented as the electricity demand and heat demand as functions of real-time electricity price. As the electricity demand and heat demand functions guarantee the optimal solution for the aggregator, the aggregator therefore can choose them as the demand bids for minimizing the energy purchase cost. Moreover, it is no longer necessary for ISO to obtain the detailed energy consumption models of aggregators, since all the parameters pertaining to energy demand and operational constraints at the customers' side are internalized in the demand bids.

The TLOP is simplified to an extended OPF problem consequently, which clears both heat and electricity and involves demand bids. The optimization problem is a joint heat and electricity energy dispatch model, involving many equality and inequality constraints. Among all the algorithms proposed for the solution of such an optimization problem, interior-point methods (IPMs) have shown good properties in terms of fast convergence and numeric robustness. In this chapter, the optimization

Table 6.1 Parameters of heating networks

Parameter	Value	Parameter	Value
C_w	4200	T_n	−9
h	0.25	$\overline{R}$	1
l	300	$\underline{R}$	0.6
T_s	85	K1/K2/K3	0.2/0.3/0.5
T_r	60	$\eta_1/\eta_2/\eta_3$	0.8/0.85/0.9

model is solved using a primal-dual interior point solver called MIPS, for Matlab Interior Point Solver, which is derived from the algorithms described in [25].

6.5 Application and Test Results

6.5.1 *Test System and Scenarios*

A test system is introduced to illustrate the technique proposed in this chapter, in which the energy exchange happens at sub-transmission networks with voltage level in a range between 30 kV and 60 kV level. The system is developed from the 30-bus system [26] with six thermal power generating units (G1–G6), three wind farms (W1–W3) and three CHP units (C1–C3). The topology diagram of the electric part is shown in Fig. 6.5a.

Customers' maximum electricity demand can be found in and it is assumed that the heat demand has the same maximum value. The capacity of each CHP unit is set as 100 MW. At each node, the customers are connected to the DH network using heating substations. Moreover, the customers are aggregated and there is one aggregator for every 500 kW customers. For instance, the total heat demand at bus 26 is 3.5 MW, hence, there are 7 aggregators at this bus. The topology diagram of the district heating system at bus 26 is shown in Fig. 6.5b.

The profiles of the hourly electricity demand (corresponding to $L^t_{e0,i} = \sum L^t_{e0,j}$), wind power potential, and heat demand (corresponding to $\sum L^t_{h0,j}$) are shown in Fig. 6.6. The electricity demand and heat demand profiles are derived from [27].

The parameters for building the demand bids of aggregators and calculating the heat losses are shown in Table 6.1.

Moreover, all the CHP units have a fixed heat-to-electricity ratio of 1.175. The economic parameters of the generating units are shown in Table 6.2, where $a/b/c$ are the coefficients of the quadratic cost functions, i.e. $C^t_{gj} = a_e(G^t_{e,gj})^2 + b_e(G^t_{e,gj}) + c_e + a_h(G^t_{h,gi})^2 + b_h(G^t_{h,gi}) + c_h$, $C^t_{gi} = a_e(G^t_{e,gi})^2 + b_e(G^t_{e,gi}) + c_e$.

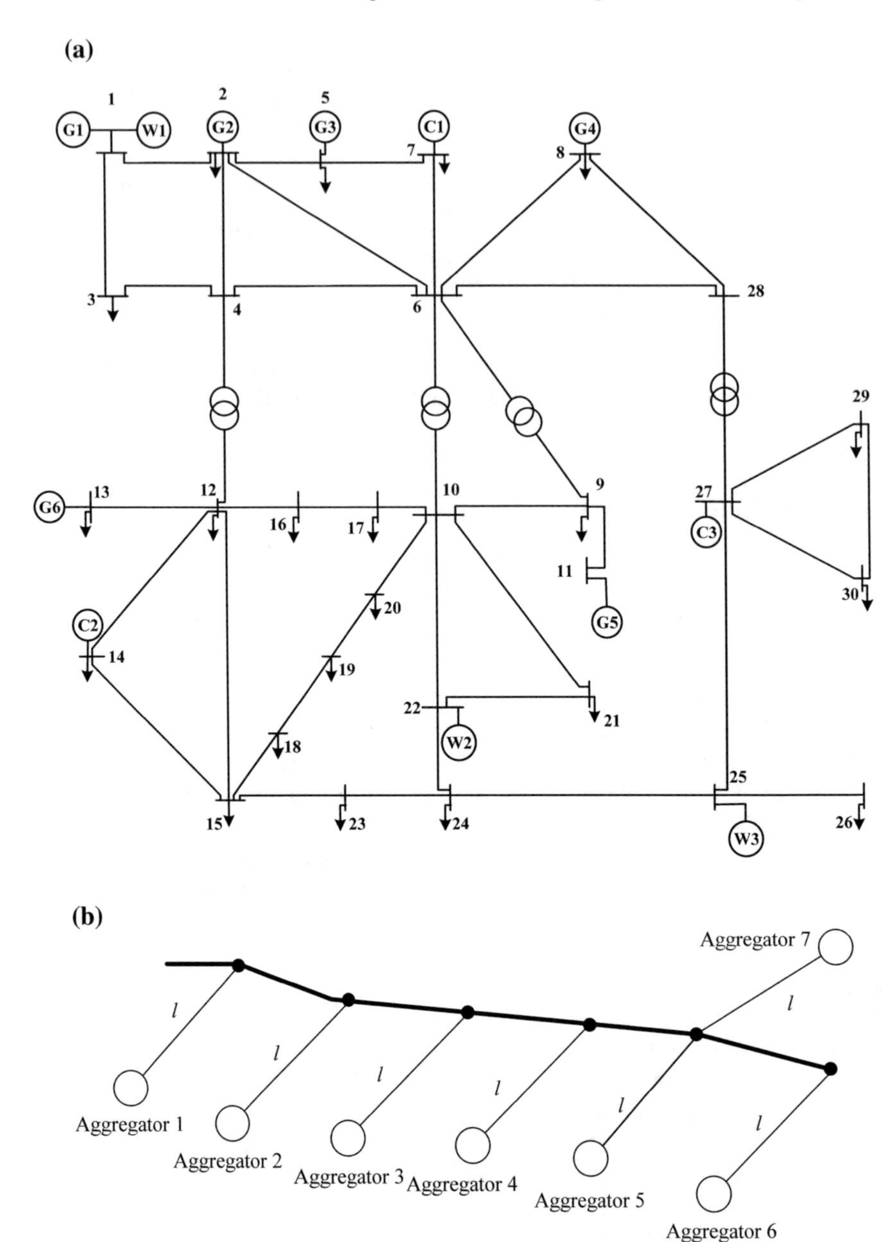

Fig. 6.5 **a** Topology diagram of the electric power system. **b** Topology diagram of the district heating system

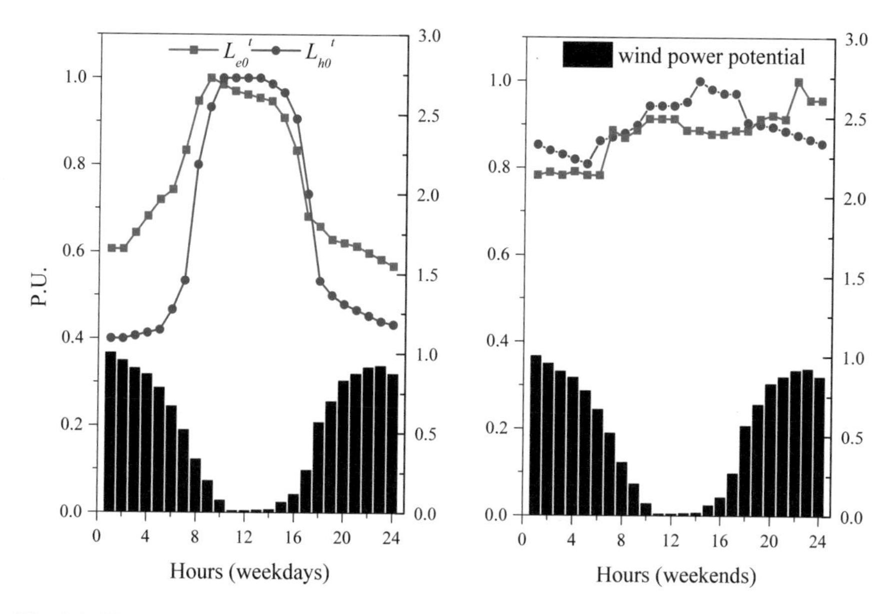

Fig. 6.6 Electricity and heat demand as well as wind power output

Table 6.2 Economy parameters of generating units

Unit	$a_e/b_e/c_e$	$a_h/b_h/c_h$
G1	0.097/50/0	–
G2	0.097/50/0	–
G3–G6	0.03/100/0	–
C1–C3	0.015/95/0	0.015/95/0
W1–W3	0	–

To illustrate the effectiveness of the proposed technique, two scenarios (S1 and S2) are modeled: in S1, the flexibility in customers' energy demand is not considered. Compared with S1, aggregators are in charge of controlling customers' energy consumption behaviors and change their energy demand in response to the supply conditions in S2. In this scenario, aggregators submit the demand bids to the ISO to participate in the centralized energy dispatch program. In other word, the aggregators' electricity demand and heat demand in this case are elastic and price dependent.

The proposed model is applied over a 24 h horizon to the test system. Moreover, it is tested on a PC with Intel 2.4 GHz 2-core processor (4 MB L3 cache), 8 GB memory. The time consumed for simulation is about 0.63 s-0.74 s. The simulation results from the two scenarios are analyzed in terms of electricity price volatility, wind power accommodation and social welfare.

6.5.2 *Simulation Results*

(a) Electricity Price Volatility

The electricity prices from the two scenarios over the 24-hour horizon are depicted in Fig. 6.7a, b, respectively. Moreover, different DR participation levels are considered, which include 10% DR (10% of the customers are involved in the DR programs), 20% DR and 30% DR cases. The impact of the different- level DR on the electricity price is shown in Fig. 6.7b. As we can see in Fig. 6.7a, the electricity prices in S1 fluctuate widely, especially in weekdays when the electricity demand waves more sharply.

Compared with the electricity price shown in Fig. 6.7a, we can see that the DR programs contribute to reducing the volatility electricity price. With the consideration of 10% DR, the electricity price fluctuation range is narrowed down to 54–98$/MWh in weekdays and 63–98$/MWh in weekends. When the DR participation level increases to 30%, the electricity price fluctuation range is further narrowed down to 54–70$/MWh in weekdays and 63–69$/MWh in weekends, which indicates that a lower volatility of the electricity price can be achieved in S2. Moreover, it can be seen that the time of peak electricity price can be reduced by the DR, and the effect can be more obvious with higher DR participation level. This is mainly due to the response of aggregators' energy demand with the electricity price. Such response of the aggregator is shown in Fig. 6.8. When the electricity is oversupplied due to the increased electricity production from the wind power, the electricity price can be relatively low and aggregator therefore may increase the electric heating ($L_{es,i}^{t}$). Then, the aggregated demand for electricity increases while the aggregated demand for heat decreases. Therefore, CHP units can decrease the amount of heat and electricity produced. The increased electricity demand along with the decreased electricity production from CHP units raises the electricity price. During the peak-hours, on the contrary, the decreased electricity demand together with the increased production of CHP units reduces the electricity price.

(b) Wind Power Accommodation

For covering both the lower wind power output scenario and higher wind power output scenario, two cases are considered. The capacity of the wind farms is set as 47 MW to model the 5% wind power penetration level in case 1, and the capacity of the farms are set as 100 MW to model the 10% wind power penetration level in case 2.

The simulation results in terms of wind power accommodation are shown in Fig. 6.9. One observation from the results is that the wind power curtailment only occurs during off-peak hours in weekdays in 5% wind power case. To meet the customers' pre-determined and constant heat demand, CHP units have to remain on and generate a certain amount of electricity, which occupies part of the proportion of wind power generation. Since the wind power output in 5% wind power case is relatively low, there is no need of curtailment in most of the time. The other observation is that a severe curtailment happens when the wind power penetration level increases to 10% in S1. Moreover, in S2, the wind power curtailment can be

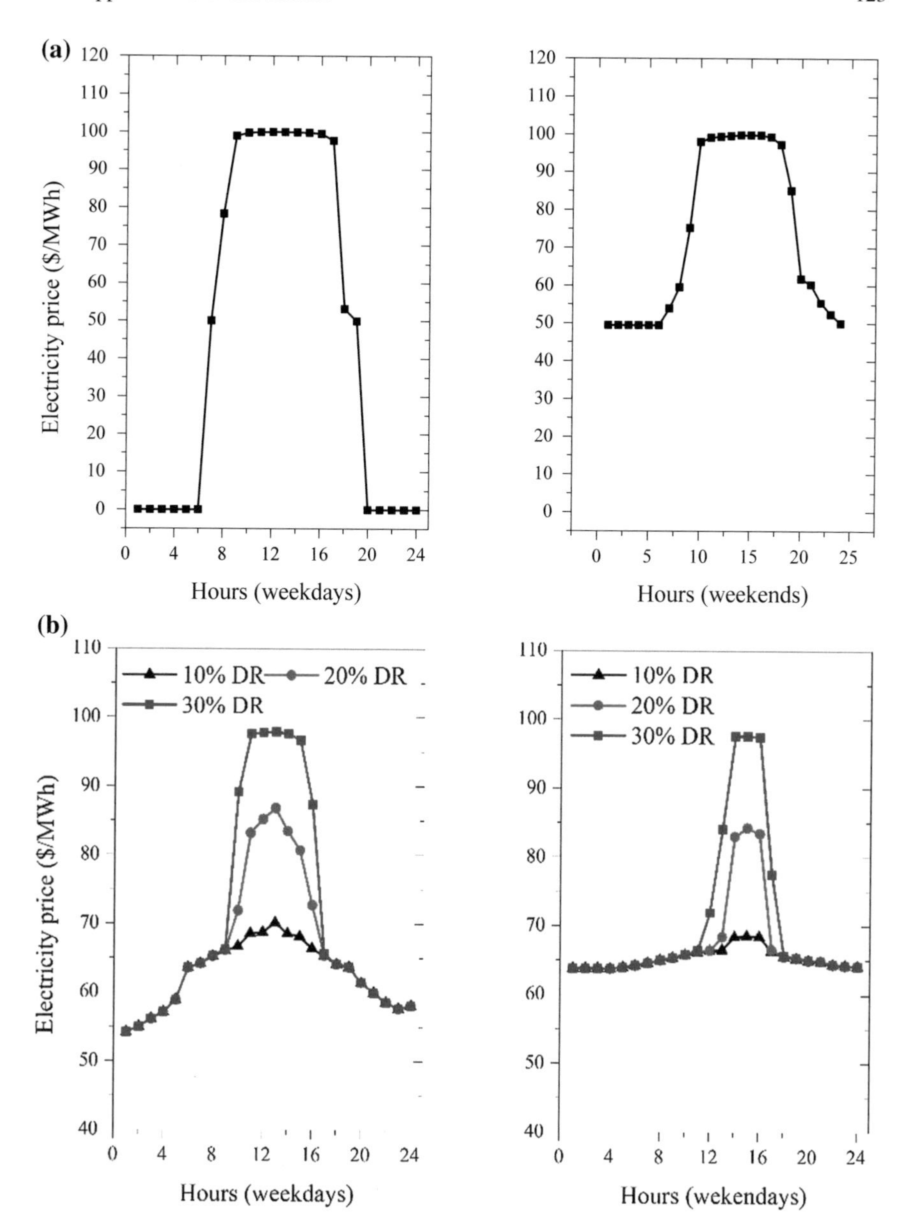

Fig. 6.7 **a** Electricity prices in S1. **b** Electricity prices in S2

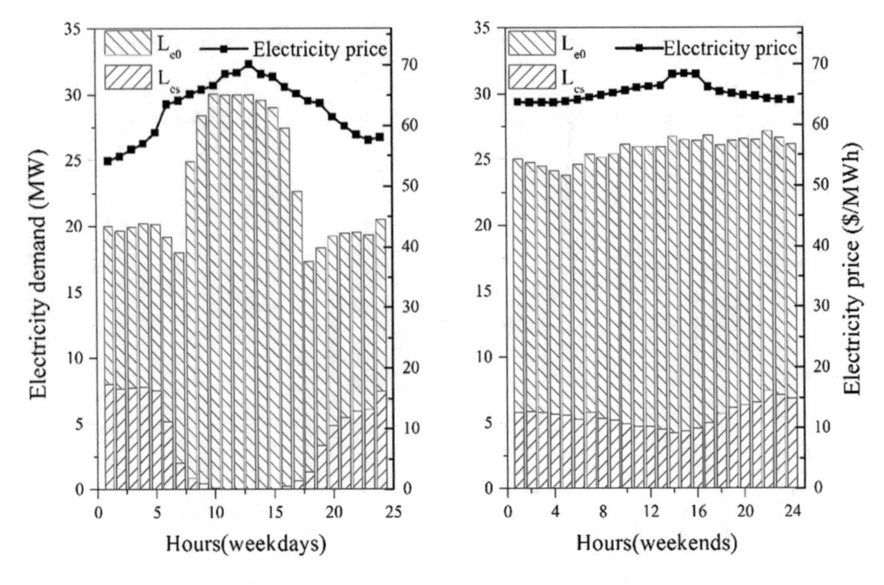

Fig. 6.8 Aggregators' response with the electricity price

Table 6.3 Social Welfare ($/h)

Scenarios		Weekdays	Weekends
S1		5207	6031
S2	10% DR	5612	6223
	20% DR	5853	6356
	30% DR	6069	6435

mitigated remarkably. During the off-peak hours, the relatively low electricity price makes customers shift the heat demand to electric heating, leading to the decrease in the electricity production from CHP units and increase in electricity demand. Thereby, the wind power output can be fully utilized.

(c) Social Welfare

The social welfare results in S1 and S2 are compared and shown in Table 6.3.

From Table 6.3, we can see that the social welfare in S2 is much higher than that in S1. With the consideration of 10% DR, the social welfare in S2 increases by 405$/h and 192$/h in weekdays and weekends, respectively. In other word, the social welfare in weekdays increases by 7.8% and the social welfare in weekends increases by 3.2% if 10% of the customers participate in the DR programs. If 30% of the customers are involved in the DR program, the social welfare in weekdays further increases by 16.6% while the social welfare in weekends further increases by 6.7%. Moreover, given that the electricity demand fluctuates more drastically in weekdays than in weekends, we can draw the conclusion that the customers' flexible resources could be more valuable when there is a high volatility of electricity demand. The total

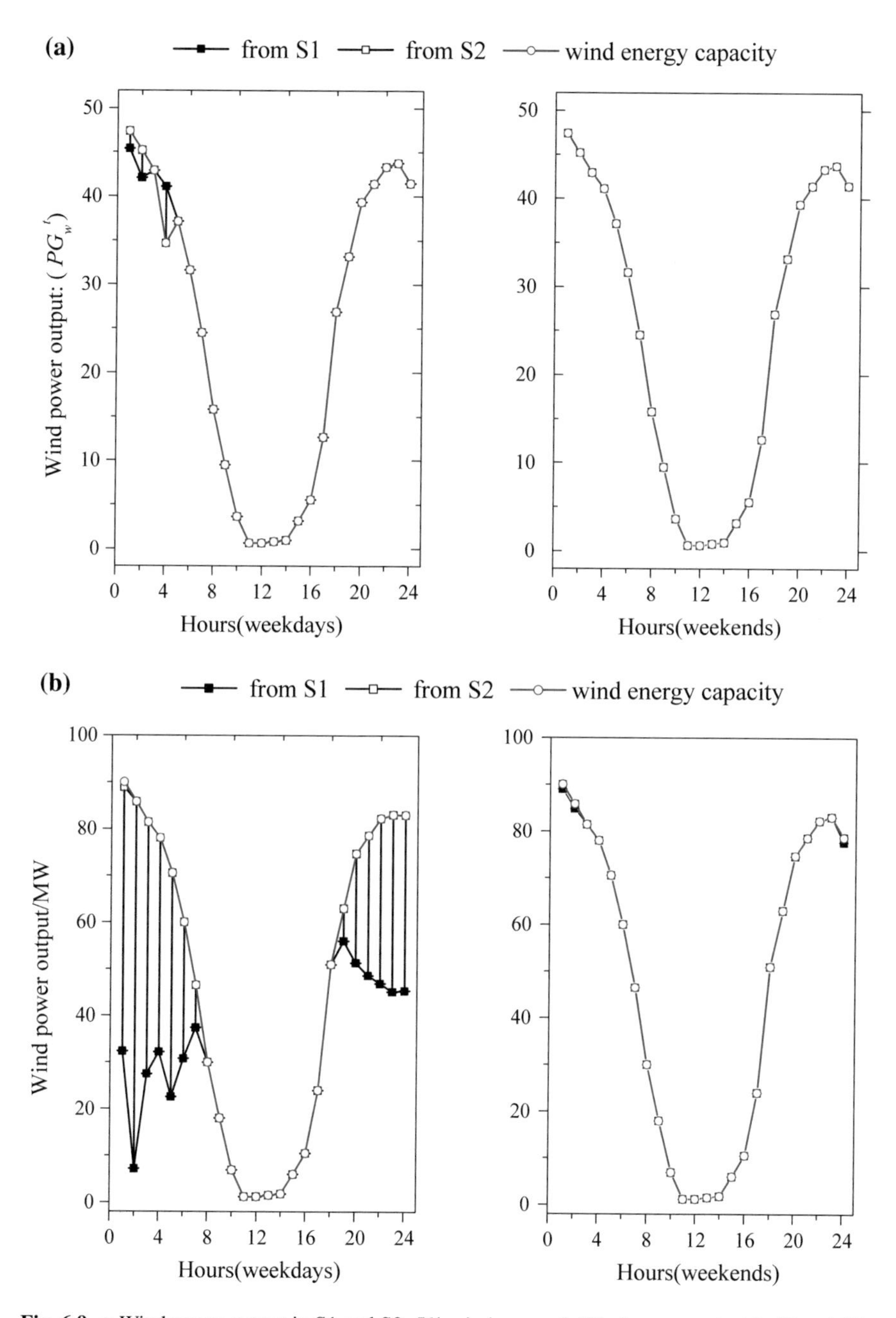

Fig. 6.9 **a** Wind power output in S1 and S2: 5% wind power. **b** Wind power output in S1 and S2: 10% wind power

social welfare increase is calculated as 753,627 $ per year with the consideration of 10% DR, assuming that only winter is considered. When the DR level increases to 20% and 30%, the total social welfare increases will further increase to 1,213,885$ and 1,601,202$, respectively. The increase of the social welfare gives reasons to encourage the DR programs and offsets the investment cost of such DR programs.

6.6 Conclusions

The inharmony between the variable wind power and the wide use of combined heat and power (CHP) has become a significant barrier to the efficient utilization of the wind power. A lot of researches have been conducted on increasing the flexibility of CHP based energy system, from the production and the network sides. This chapter investigates the utilization of the customers' flexible energy demand to provide additional balancing resources and release the inharmony between the variable wind power and the wide use of CHP. We find that the integration of heat and electricity systems provides multiple options to customers for fulfilling their energy demand. The built-in flexibility in customers' energy demand provides desirable flexible resources for maintaining the balance between energy supply and demand and achieving the efficient utilization of the wind power.

References

1. B. Zeng, J. Zhang, X. Yang, J. Wang, J. Dong, Y. Zhang, Integrated planning for transition to low-carbon distribution system with renewable energy generation and demand response. IEEE Trans. Power Syst. **29**, 1153–1165 (2014)
2. G. Streckienė, V. Martinaitis, A.N. Andersen, J. Katz, Feasibility of CHP-plants with thermal stores in the German spot market. Appl. Energy **86**, 2308–2316 (2009)
3. P.F. Bach, Towards 50% wind electricity in Denmark: Dilemmas and challenges. Eur. Phys. J. Plus **131**, 1–12 (2016)
4. X. Lu, M.B. McElroy, W. Peng, S. Liu, C.P. Nielsen, H. Wang, Challenges faced by China compared with the US in developing wind power. Nat. Energy **1**, 16061, 05/23/online (2016)
5. M.G. Nielsen, J.M. Morales, M. Zugno, T.E. Pedersen, H. Madsen, Economic valuation of heat pumps and electric boilers in the Danish energy system. Appl. Energy **167**, 189–200 (2015)
6. T. Jónsson, P. Pinson, H. Madsen, On the market impact of wind energy forecasts. Energy Econ. **32**, 313–320 (2010)
7. X. Chen, C. Kang, M.O. Malley, Q. Xia, J. Bai, C. Liu et al., Increasing the flexibility of combined heat and power for wind power integration in china: modeling and implications. IEEE Trans. Power Syst. **30**, 1848–1857 (2015)
8. A. Fragaki, A.N. Andersen, Conditions for aggregation of CHP plants in the UK electricity market and exploration of plant size. Appl. Energy **88**, 3930–3940 (2011)
9. Z. Li, W. Wu, J. Wang, B. Zhang, T. Zheng, Transmission-constrained unit commitment considering combined electricity and district heating networks. IEEE Trans. Sustain. Energy **7**, 480–492 (2016)

10. J. Duquette, A. Rowe, P. Wild, J. Yan, Thermal performance of a steady state physical pipe model for simulating district heating grids with variable flow. Appl. Energy **178**, 383–393 (2016)
11. X. Zhang, M. Shahidehpour, A. Alabdulwahab, A. Abusorrah, Optimal expansion planning of energy hub with multiple energy infrastructures. IEEE Trans. Smart Grid **6**, 2302–2311 (2015)
12. M. Moeini-Aghtaie, P. Dehghanian, M. Fotuhi-Firuzabad, A. Abbaspour, Multiagent genetic algorithm: an online probabilistic view on economic dispatch of energy hubs constrained by wind availability. IEEE Trans. Sustain. Energy **5**, 699–708 (2014)
13. J. Sjödin, D. Henning, Calculating the marginal costs of a district-heating utility. Appl. Energy **78**, 1–18 (2004)
14. S. Bahrami, A. Sheikhi, From demand response in smart grid toward integrated demand response in smart energy hub. IEEE Trans. Smart Grid **7**, 650–658 (2016)
15. A. Sheikhi, M. Rayati, S. Bahrami, A.M. Ranjbar, Integrated demand side management game in smart energy hubs. IEEE Trans. Smart Grid **6**, 675–683 (2015)
16. C. Shao, Y. Ding, J. Wang, Y. Song, Modeling and integration of flexible demand in heat and electricity integrated energy system. IEEE Trans. Sustain. Energ. 9(1):361–370 (2018)
17. P. Mancarella, G. Chicco, Real-time demand response from energy shifting in distributed multi-generation. Smart Grid IEEE Trans. **4**, 1928–1938 (2013)
18. Pirouti and Marouf, Modelling and analysis of a district heating network. Cardiff University (2013)
19. J. Gustafsson, J. Delsing, J.V. Deventer, Improved district heating substation efficiency with a new control strategy. Appl. Energy **87**, 1996–2004 (2010)
20. J. Duquette, A. Rowe, P. Wild, Thermal performance of a steady state physical pipe model for simulating district heating grids with variable flow. Appl. Energy **178**, 383–393 (2016)
21. X.S. Jiang, Z.X. Jing, Y.Z. Li, Q.H. Wu, W.H. Tang, Modelling and operation optimization of an integrated energy based direct district water-heating system. Energy **64**, 375–388 (2013)
22. B. Rolfsman, Combined heat-and-power plants and district heating in a deregulated electricity market. Appl. Energy **78**, 37–52 (2004)
23. C. Ruiz, A.J. Conejo, Pool strategy of a producer with endogenous formation of locational marginal prices. Power Syst. IEEE Trans. **24**, 1855–1866 (2009)
24. W. Yu-Chi, A.S. Debs, R.E. Marsten, A direct nonlinear predictor-corrector primal-dual interior point algorithm for optimal power flows. IEEE Trans. Power Syst. **9**, 876–883 (1994)
25. J.D. Weber, T.J. Overbye, A two-level optimization problem for analysis of market bidding strategies, in *Power Engineering Society Summer Meeting*, vol. 2, pp. 682–687 (1999)
26. H. Wang, C.E. Murillo-Sánchez, R.D. Zimmerman, R.J. Thomas, On Computational Issues of market-based optimal power flow. IEEE Trans. on Power Syst. **22**, 1185–1193 (2007)
27. C. Wang, M.H. Nehrir, Analytical approaches for optimal placement of distributed generation sources in power systems. IEEE Trans. Power Syst. **19**, 2068–2076 (2004)

Chapter 7
A Three-Level Framwork for Utilizing the Demand Response to Improve the Operation of the Integrated Energy Systems

7.1 Introduction

Chaper 6 has analyzed the demand response potential of cutomers (usually refer to buildings) in the distribution-level heat and electricity integrated energy system. This chapter proposes a framework for utilizing the demand response to improve the operation of the integrated energys sytem (IES) which has gained rapid development recently [1]. The framework involves three levels of the integrated energy system: aggregation of the smart buildings, distribution system, and transmission system or sub-transmission system. In the framework, the buildings' demand response potential can be fully utilized and the operational flexibility of the transmission-level integrated energy system can be significantly imptoved.

As discussed, the development of CHP based HE-IES contributes to the energy saving and emissions reduction [2, 3]. However, the coupling between different energy vectors may limit the operational flexibility of the system, which becomes a barrier to the integration of variable wind power [4]. As we know, the output of wind power is variable and uncertain. Hence, the other generation must be flexible and able to provide the balancing power to absorb the variable wind power. However, since the electricity generation of CHP units is constrained by their heat production which must target customers' heat demand, it is difficult for the CHP units to adjust their electricity output frequently to provide the required balancing power. Therefore, the efficient utilization of the wind power is difficult to be achieved in such CHP-based HE-IES system [3]. In Northern and Northeastern provinces of China, for example, the wind power curtailment rate has exceeded 20% [5]. Therefore, additional flexibility is required to provide the balancing power for integrating the variable wind power.

Recently, demand response (DR) has become one of the most preferred options for providing the balancing power. Among all the different flexible demand-side resources, smart buildings are expected to play a significant role in such DR programs [6]. With the development of the DR-enabled technologies, smart buildings are able to modify their energy consumption behaviors so as to provide the balancing power in

Y. Ding et al., *Integration of Air Conditioning and Heating into Modern Power Systems*,
https://doi.org/10.1007/978-981-13-6420-4_7

response to supply conditions [7]. Many studies have been conducted on the energy management methods and DR strategies to leverage the demand flexibility of the buildings [8–10]. In [8], it was illustrated that the commercial buildings' heating ventilation and air conditioning (HVAC) systems can be a valuable resource for the frequency regulation. In [9], the authors proposed a reward based DR strategy for residential consumers to provide the electric balancing power so as to shave network peaks. In [10], the authors considered a smart building operator that is capable of modulating the building' energy consumption via price signals to provide the balancing power that an independent system operator (ISO) requires. To facilitate the integration and utilization of such demand-side resources, a demand response market (DRX) concept has been proposed [11–13], where the demand-side resources are stimulated to provide the balancing power by the incentive related to energy prices. Moreover, an iteration-based method have been proposed to clear the DRX market [12, 13]. In the iteration-based method, DR providers adjust the amount of the balancing power they want to provide in response to the prices which are adjusted in turn by the market operator. This process is repeated iteratively until the market equilibrium (overall optimization) is obtained.

As introduced above, there are some researches regarding the DR of smart buildings. However, very few studies have been conducted on the DR applied to a heat and electricity integrated energy system (HE-IES). The DR applied to buildings in the HE-IES is quite different from the traditional DR in the electric power system. Hence, the above energy management methods and market framework might not be proper for the DR in the HE-IES. Firstly, those DR management methods mainly focus on encouraging buildings to adjust their electricity demand so as to provide the electric balancing power. However, heat balancing power is also necessary to fully integrate the variable wind power in the HE-IES. Given that the operation of CHP units is constrained by their heat output corresponding to customers' heat demand, encouraging buildings to adjust their heat demand in response to the supply conditions is more crucial for relaxing the production constraints of CHP units and to increase the system flexibility. Secondly, the DR control strategies applied in the existing studies are limited to load shifting taking advantage of the thermal inertia of the buildings [8–10]. However, energy substitution referring to switching the sources of the consumed energies is also a practicable DR control strategy in the HE-IES context, considering the energy conversion between different heat power and electric power. Thirdly, to fully utilize the buildings' demand flexibility in the HE-IES, the existing DRX market should also be modified. The traditional DRX market is typically incorporated into the day-ahead energy market clearing [11, 13]. However, during the day-ahead energy market, it is difficult to exactly determine how much balancing power will be required during the next-day operational hour, considering that the wind power output is difficult to predict. Moreover, the very short-term power fluctuation from wind energy cannot be balanced using DR, which may lead to price spikes and higher energy generation cost. Therefore, a real-time DRX is necessary. Additionally, although the convergence of the iteration-based method can be guaranteed, it may require many iterations in the calculation when there are many market participants. Hence, a market clearing technique more efficient than the conventional iteration-based clearing method should be developed to meet the higher requirement for the fast clearing in the real-time market.

Based on the above analysis, this chapter expands the buildings' DR to the HE-IES. The utilization of smart buildings to provide the required heat and electric balancing powers is investigated. Comprehensive demand flexibility of buildings is firstly exploited by combining load shifting and energy substitution strategies in this chapter. Then, an aggregation method is proposed to aggregate multiple buildings' flexible energy demands. Moreover, a real-time DRX market is also developed where the building aggregators are stimulated to adjust buildings' energy consumption behaviors and provide the required balancing power. It has been demonstrated that the real-time DRX market is able to balance the very short-term power fluctuation from wind energy through DR, which helps to reduce the price spike and to reduce the generation costs. Additionally, a novel optimum feasible region method is proposed to achieve the fast clearing of the DRX market, which is more applicable for the real-time market. This chapter includes research related to the framework for incorporating DR of smart buildings into the integrated heat and electricity energy system by [14].

7.2 Energy Demand of Smart Buildings

As a typical facility that can be considered as an energy hub, smart buildings can be identified as a unit that involves the conversion, consumption and storage of different energy carriers [15]. The schematic diagram of the smart buildings' energy demands in the HE-IES context is illustrated in Fig. 7.1. As shown in Fig. 7.1, the energy demands of the building can be divided into: heating demand and non-heating demand. Heating demand refers to the need of space-heating service, and non-heating demand includes the lighting, pumps, fans and other electric appliances. Moreover, non-heating demand is met through the electric power from the electricity network, and the heating demand can be met through the heat power from the district heating (DH) network or the electric heating.

Such buildings are able to participate actively in the DR programs. Control strategies applied in DR programs can be classified into three categories: load curtailment, load shifting, and substitution [16]. Load curtailment is usually not considered since it usually comes at the cost of comfort. The load shifting is utilized in this chapter. Moreover, the storage devices are not considered as shown in Fig. 7.1. Rather, the purpose is to utilize the buildings' thermal inertia to provide the thermal storage in this chapter [17]. Energy substitution referring to switching the source of the consumed energy is also practicable in this context since the electric power can be converted to the heat power by using the electric heating devices.

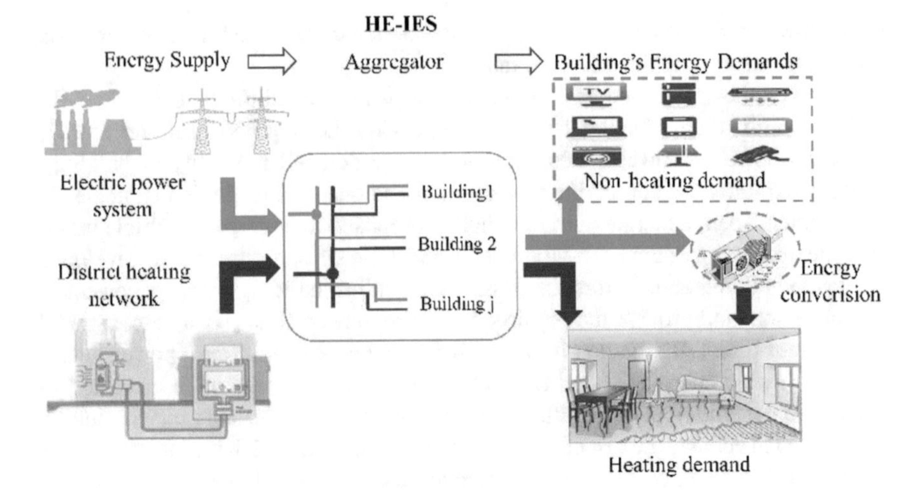

Fig. 7.1 Schematic diagram of the smart buildings' energy demand

7.2.1 *Modeling Individual Building's Energy Demand Based on the Comprehensive DR Strategy*

Energy demands of a building are defined as the demands for the imported heat power ($L_{\mathrm{H},IMP}^{bui}$) and electric power ($L_{E,IMP}^{bui}$) from the energy networks. This section develops the energy demand model to describe their expressions. Firstly, energy substitution strategy, which is not considered in most studies is developed to exploit the buildings' demand flexibility. In the energy substitution strategy, buildings have two options to fulfill their space-heating needs: they can directly use the heat power produced by the CHP units, or convert the electric power into the heat power by using electric heating devices. Hence, the total value of the heat injected into the building can be calculated as the following:

$$L_{H,\mathrm{ALL}}^{bui} = L_{H,ES}^{bui} + L_{H,IMP}^{bui} \tag{7.1}$$

Moreover, $L_{H,ES}^{bui}$ and $L_{H,IMP}^{bui}$ are expressed as:

$$L_{H,ES}^{bui} = \eta_{ES} L_{E,ES}^{bui},\ L_{H,IMP}^{bui} = m_j c_w (\tau_{s,j} - \tau_{r,j}) \tag{7.2}$$

The bounds of the mass flow m_j is given by:

$$m_j^{\min} \le m_j \le m_j^{\max} \tag{7.3}$$

Moreover, the load shifting strategy based on the buildings' thermal inertia is also considered to increase the demand flexibility of buildings. In this chapter, the simulation model developed to study the thermal dynamics of a building is based on the equivalent thermal parameter (ETP) method [18]. The ETP model can be described as follows:

$$\dot{\tau}_a = \frac{1}{c_a}(-u_{a-h}(\tau_a - \tau_h) - u_{a-o}(\tau_a - \tau_o) + L_{\mathrm{H},ALL}^{bui})$$
$$\dot{\tau}_h = \frac{1}{c_h}(u_{a-h}(\tau_a - \tau_h) - u_{h-o}(\tau_h - \tau_o)) \tag{7.4}$$

Moreover, to ensure no loss of comfort in the building, (7.5) is enforced:

$$\tau_a^{\min} \leq \tau_a^t \leq \tau_a^{\max} \tag{7.5}$$

where $\tau_a^{\min}$ and $\tau_a^{\max}$ are the thresholds of the comfortable temperature.

The building's flexible energy demands based on the comprehensive DR strategy therefore can be expressed as:

$$\begin{aligned} L_{E,IMP}^{bui} &= L_{E,ES}^{bui} + L_{\mathrm{E,E0}}^{\mathrm{bui}};\, L_{H,IMP}^{bui} = L_{H,\mathrm{ALL}}^{bui} - L_{H,ES}^{bui} \\ \tau_a^{t+1} - \tau_a^t &= \frac{\Delta t}{c_a}(-u_{a-h}(\tau_a - \tau_h) - u_{a-o}(\tau_a - \tau_o) + L_{\mathrm{H},ALL}^{bui}) \\ \tau_h^{t+1} - \tau_h^t &= \frac{\Delta t}{c_h}(u_{a-h}(\tau_a - \tau_h) - u_{h-o}(\tau_h - \tau_o)) \\ \tau_a^{\min} &\leq \tau_a^t \leq \tau_a^{\max} \end{aligned} \tag{7.6}$$

It should be noted that the thermal dynamic model as (7.4) has been transformed into a discrete difference equation.

7.2.2 Energy Demand Aggregation of Multiple Buildings

The capacity of a single building's energy demand is limited, which means it is difficult for a single building to directly participate in the energy market. Hence, the building aggregators as an independent entity is introduced to represent a cluster of buildings. The electric power demand of the aggregator is equal to the sum of the electricity demand of all the buildings it serves, which is expressed as:

$$P_{E,i}^{dem} = \sum\nolimits_{j \in \Xi_i} L_{E,IMP,j}^{bui} \tag{7.7}$$

The heat demand of the aggregator can be calculated as:

$$P_{H,i}^{dem} = m_i c_w(\tau_{s,i} - \tau_{r,i}) = \sum\nolimits_{j \in \Xi_i} L_{H,IMP,j}^{bui} + P_{H,i}^{loss} \tag{7.8}$$

where $P_{H,i}^{loss}$ is the heat loss which can be expressed as:

$$P_{H,i}^{loss} = m_i c_w(\tau_{s,i} - \tau_{r,i}) - \sum\nolimits_j m_j c_w(\tau_{s,j} - \tau_{r,j}) \tag{7.9}$$

The aggregators' gross water flow is expressed as:

$$m_i = \sum\nolimits_{j \in \Xi_i} m_j \tag{7.10}$$

In addition, the temperature drops exponentially when water flows in pipes [19], hence:

$$\begin{aligned} \tau_{s,j} &= (\tau_{s,i} - \tau_o)e^{-hl/c_w m_i} + \tau_o \\ \tau_{r,j} &= (\tau_{r,i} - \tau_o)e^{hl/c_w m_i} + \tau_o \end{aligned} \tag{7.11}$$

Using the equivalent infinitesimal replacement, (7.11) can be approximately written as:

$$\begin{aligned} \tau_{s,j} &= (\tau_{s,i} - \tau_o)\left(1 - hl/c_w m_i\right) + \tau_o \\ \tau_{r,j} &= (\tau_{r,i} - \tau_o)\left(1 + hl/c_w m_i\right) + \tau_o \end{aligned} \tag{7.12}$$

Combining (7.10) and (7.12), the heat loss can be expressed as:

$$P_{H,i}^{loss} = (\tau_{s,i} + \tau_{r,i} - 2\tau_o)hl \tag{7.13}$$

In summary, the energy demands of the aggregator can be expressed as:

$$\begin{aligned} P_{E,i}^{dem} &= \sum\nolimits_{j \in \Xi_i} L_{E,IMP,j}^{bui} \\ P_{H,i}^{dem} &= \sum\nolimits_{j \in \Xi_i} L_{H,DH,j}^{bui} + P_{H,i}^{loss} \end{aligned} \tag{7.14}$$

while the heat loss $P_{H,i}^{loss}$ is independent on m_j,m_i.

In this way, the aggregated energy demands can be represented as the linear functions of the individual energy demands, which is very helpful in simplifying the further calculation.

7.3 Concept and Framework of the Real-Time DRX Market

To fully realize the potential of the demand flexibility in providing the required balancing power, the energy market needs to be tailored to incorporate the participation of the buildings' aggregators and fair incentives should be provided [20]. Recently, the DRX market concept has been proposed where the DR offers are provided and cleared to improve the energy system operation. However, the traditional DRX market, which is typically incorporated into the day-ahead energy market clearing, is not able to balance the very short-term wind power fluctuation. Thus, a novel real-time DRX is developed in this chapter. This section introduces the market framework, optimization models and clearing of the proposed real-time DRX market.

7.3.1 Three-Level Framework of the DRX Market

The main idea of the market mechanism is illustrated in Fig. 7.2. As shown in Fig. 7.2, the framework integrates DR and system optimization across three levels of the integrated energy system: aggregation of the smart buildings, distribution system, and transmission system or sub-transmission system – through the interactions of three key entities, i.e., the building aggregators, the DSOs and the ISO. The responsibilities of the three entities in the DRX market are summarized as:

(a) ISO: ISO runs the integrated heat and electricity dispatch (IHED) model with the consideration of the flexible energy demands. Running the IHED model, the DR schedules, i.e. how much balancing power is required at each bus for each time period, can be determined. Then, the target values about the desired balancing power is sent to the DSOs.
(b) DSOs: each DSO is responsible for each bus in the transmission system. Once knowing how much balancing power is required, the DSOs runs a bi-level optimization model and determines the proper energy prices which can stimulate the building aggregators to provide the balancing power corresponding to the target values. Then, the energy prices are sent out to the aggregators.
(c) Aggregators: each aggregator represents a cluster of buildings. Aggregators can adjust the buildings' energy consumption behaviors to provide the balancing power in response to the energy prices.

An overall timeline of the market is shown in Fig. 7.2b. At the beginning of the operation hour (08:00 to 09:00), the real-time DRX market is launched. In the DRX market, the ISO firstly evaluates how much balancing power is required from the DRX market for the next several intervals during the operation hour. Then, the DSOs stimulate the aggregators to respond to the balancing power requirement through the price signals. During operation, the aggregators utilize energy management systems of the building to modify the buildings' energy consumption behaviors for providing the scheduled balancing power.

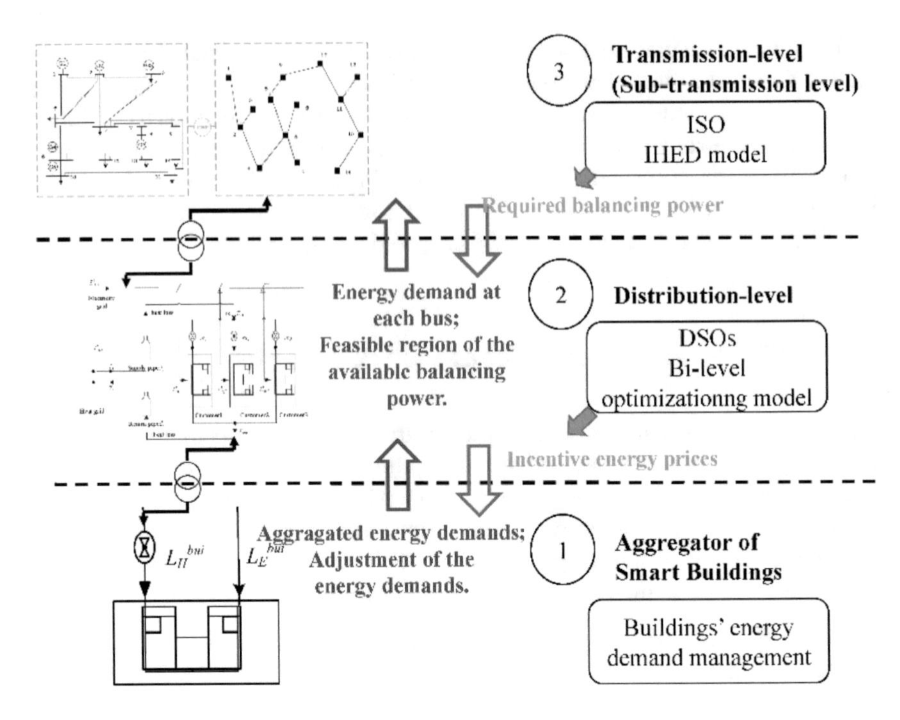

(a) The three-level market framework

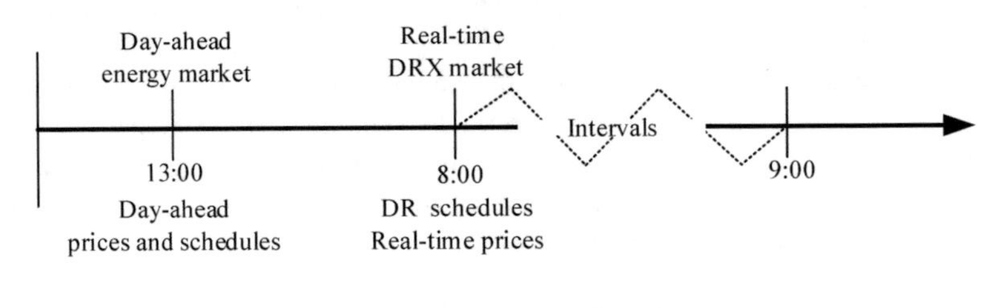

(b) Time line

Fig. 7.2 Market framework of the real-time DRX market **a** The three-level market framework **b** Time line

7.3.2 *Optimization Models in the DRX Market*

(a) ISO's Strategy

In the DRX market, the ISO firstly runs the IHED model to determine the scheduling plan of the generating units and buildings aggregator for the next few time intervals. The familiar methods for the combined optimization of electric power and heat power can be found in [21, 22]. The formulation of the IHED model is shown in (7.15)–(7.27).

The objective of the dispatch model is to maximize the social welfare, as shown in (7.16). Besides the real-time generation outputs ($P_{E,\phi}^{gen1}$, $P_{H,\phi}^{gen1}$, $P_{E,\varphi}^{gen1}$), the variables to be optimized are the desired balancing power ($\Delta P_{E,i}^{req}$, $\Delta P_{H,i}^{req}$) as the difference between the before-DR energy demands ($P_{E,i}^{dem0}$, $P_{H,i}^{dem0}$)and the real-time and after-DR energy demands ($P_{E,i}^{dem1}$, $P_{H,i}^{dem1}$). It should be noted that the subscripts t for all variables are omitted for simplicity.

$$Max\ F = \sum_{t=1}^{NT_H} \Big(\sum\nolimits_{i=1}^{N_A} R_i(P_{E,i}^{dem0} + \Delta P_{E,i}^{req}, P_{H,i}^{dem0} + \Delta P_{H,i}^{req}) -$$

$$\cdots \sum\nolimits_{\phi=1}^{N_\phi} C_\phi^{gen}(P_{E,\phi}^{gen1}, P_{H,\phi}^{gen1}) - \sum\nolimits_{\varphi=1}^{N_\varphi} C_\varphi^{gen}(P_{E,\varphi}^{gen1}) \Big) \quad (7.15)$$

In (7.16), the first part denotes the total benefits of the consumers, while the second part and the third part denote the generation cost of the electricity-only generating units and the generation cost of the CHP units, respectively. It should be noted that the generation cost of the wind power is set to zero here.

The electric power and heat power balance constraints are expressed as (7.16).

$$\sum\nolimits_{\varphi\in\Psi_n} P_{E,\varphi}^{gen1} + \sum\nolimits_{\phi\in\Psi_n} P_{E,\phi}^{gen1} + \sum\nolimits_{w\in\Psi_n} P_{E,w}^{gen1} - \sum\nolimits_{i\in\Psi_n} \left(P_{E,i}^{dem0} + \Delta P_{E,i}^{req}\right)$$

$$= \sum\nolimits_{p\in\Theta_n} V_n V_p \left[G_{np}\cos(\delta_n - \delta_p) + B_{np} sin(\delta_n - \delta_p)\right]$$

$$\sum\nolimits_{\varphi\in\Psi_n} P_{E,\varphi}^{gen1} + \sum\nolimits_{\phi\in\Psi_n} P_{E,\phi}^{gen1} - \sum\nolimits_{i\in\Psi_n} \left(P_{H,i}^{dem0} + \Delta P_{H,i}^{req}\right) = \sum\nolimits_{p\in\Psi_n} P_{H,np}^{line} \quad (7.16)$$

where $\varphi \in \Psi_n$ identifies the thermal power generating unit located at bus n and $p \in \Theta_n$ identifies the bus p connected to bus n.

The amount of heat power through a pipeline $P_{H,np}^{line}$ is calculated by:

$$P_{H,np}^{line} = c_w m_{np} \left(\tau_{s,n} - \tau_{s,p}\right) \quad (7.17)$$

The static pressure drop of a pipe is directly proportional to the square of its mass flow rate. The pressure drop is given by (7.19) where k_{np} is a constant.

$$\rho_n - \rho_p = k_{np} m_{np}^2 \quad (7.18)$$

The supply and return networks are modelled as two independent networks coupled only through flow rates [22]. Moreover, the temperature drops exponentially:

$$\begin{aligned} \tau_{s,p} &= (\tau_{s,n} - \tau_a)(1 - hl_{s,np} / c_w m_{np}^t) + \tau_a \\ \tau_{r,p} &= (\tau_{r,n} - \tau_a)(1 + hl_{r,np} / c_w m_{np}^t) + \tau_a \end{aligned} \tag{7.19}$$

Equations (7.20)–(7.24) represent the power output constraints for the generating units.

$$P_{E,\varphi}^{\min} \le P_{E,\varphi}^{gen1} \le P_{E,\varphi}^{\max} \quad \forall \varphi \tag{7.20}$$

$$P_{E,w}^{\min} \le P_{E,w}^{gen1} \le P_{E,w}^{\max} \quad \forall w \tag{7.21}$$

$$P_{E,\phi}^{\min} \le P_{E,\phi}^{gen1} \le P_{E,\phi}^{\max} \quad \forall \phi \tag{7.22}$$

$$P_{H,\phi}^{\min} \le P_{H,\phi}^{gen1} \le P_{H,\phi}^{\max} \quad \forall \phi \tag{7.23}$$

$$P_{H,\phi}^{gen1} = \gamma_{HE,\phi} P_{E,\phi}^{gen1} \tag{7.24}$$

The transmission capacity limits are expressed as:

$$-T_{np}^{\max} \le T_{np} \le T_{np}^{\max}, \quad \forall p \in \Theta_n \tag{7.25}$$

The phase angle bounds for each electric bus and the static pressure bounds for each DH node are:

$$\delta_n^{\min} \le \delta_n \le \delta_n^{\max} \quad \forall n \tag{7.26}$$

$$\rho_n^{\min} \le \rho_n \le \rho_n^{\max} \quad \forall n \tag{7.27}$$

In this chapter, both the heat and electricity are priced based on the LMP method. LMP measures the least cost to supply an additional unit of load at that location from the resources of the system. Moreover, it has been revealed that the LMPs are the shadow prices of the power balance equality constraints of the dispatch model [23].

(b) DSOs' Strategy

Once knowing how much balancing power is required, the DSOs will encourage the aggregators to respond to the balancing power requirement by adjusting buildings' energy consumption behaviors. From the perspective of the DSOs, each aggregator can be treated as a profit-driven and independent agent; therefore, an appropriate amount of balancing power can be provided by the aggregators if proper incentive prices are offered to them. The DSOs' strategies are modelled as an optimization problem which minimizes the deviations between the desired balancing power and the actual balancing power provided from the DR. Moreover, the problem can be formulated as a bi-level optimization model, where the lower-level problem (LP) represents the aggregator's optimal strategy, and the higher-level problem (HP) minimizes the deviation between the desired ($\Delta P_{E,n}^{req}$, $\Delta P_{H,n}^{req}$) balancing power and actual balancing power provided by the DRX ($\Delta P_{E,n}^{avi}$, $\Delta P_{H,n}^{avi}$).

The bi-level problem is formulated as:

HP:

$$\begin{aligned} &Min \sum\nolimits_{t=1}^{NT_H} \left(\varepsilon_{E,n}\left|\Delta P_{E,n}^{avi} - \Delta P_{E,n}^{req}\right| + \varepsilon_{H,n}\left|\Delta P_{H,n}^{avi} - \Delta P_{H,n}^{req}\right|\right) \\ &s.t.\ \Delta P_{E,n}^{avi} = \sum\nolimits_{i\in\Psi_n} \Delta P_{E,i}^{avi} = \sum\nolimits_{i\in\Psi_n} P_{E,i}^{dem} - P_{E,i}^{dem0} \\ &\Delta P_{H,n}^{avi} = \sum\nolimits_{i\in\Psi_n} \Delta P_{H,i}^{avi} = \sum\nolimits_{i\in\Psi_n} P_{H,i}^{dem} - P_{H,i}^{dem0} \\ &P_{E,i}^{dem}, P_{H,i}^{dem} \in \text{LP} \end{aligned} \tag{7.28}$$

LP:

$$Max \sum\nolimits_{t=1}^{N_T} \left(R_i(P_{E,i}^{dem}, P_{H,i}^{dem}) - p_e P_{E,i}^{dem} - p_h P_{H,i}^{dem}\right) \tag{7.29}$$

$$\begin{aligned} P_{E,i}^{dem} &= \sum\nolimits_{j\in\Xi_i} L_{E,IMP,j}^{bui} \\ P_{H,i}^{dem} &= m_i C_w (T_{s,i} - T_{r,i}) = L_{H,IMP,j}^{bui} / \left(1 - \nu_{loss}\right) \\ m_i &= \sum\nolimits_{j\in\Xi_i} m_j \end{aligned} \tag{7.30}$$

$$L_{E,\text{IMP},j}^{bui} = L_{E,ES,j}^{bui} + L_{E,E0,j}^{bui},\ L_{H,ALL,j}^{bui} = L_{H,ES,j}^{bui} + L_{H,\text{IMP},j}^{bui} \tag{7.31}$$

$$\left(\begin{aligned} \tau_a^{t+1} - \tau_a^t &= \frac{\Delta t}{c_a}(-u_{a-h}(\tau_a - \tau_h) - u_{a-o}(\tau_a - \tau_o) + L_{H,\text{ALL}}^{bui}) \\ \tau_h^{t+1} - \tau_h^t &= \frac{\Delta t}{c_h}(u_{a-h}(\tau_a - \tau_h) - u_{h-o}(\tau_h - \tau_o))\forall j \\ \tau_a^{\min} &\le \tau_a^t \le \tau_a^{\max} \end{aligned}\right) \tag{7.32}$$

In (7.28), $\varepsilon_{E,n}$ and $\varepsilon_{H,n}$ are the weighted factors which are positively correlated to the absolute value of $\Delta P_{E,n}^{req}$ and $\Delta P_{H,n}^{req}$, respectively. Equations (7.29)–(7.32) represent the strategy of the aggregator for maximizing its benefit by controlling the buildings energy consumption behaviors in response to the energy prices. Equation (7.29) denotes the relationships between the aggregator's energy demands and the downstream demands. Equation (7.31) stands for the power balance in the building.

The bi-level optimization problem can be converted to a standard optimization problem by replacing the low-level problems by their Karash-Kuhn-Tucker (KKT) optimality condition. Then, the problem can be solved using the primal-dual interior point method [24].

7.3.3 Clearing of the DRX Market

As introduced above, the basic processes in the DRX market are: the ISO firstly determines how much balancing power is required; then, the DSOs run the bi-level optimization model to determine the incentive energy prices which are sent out to the aggregators; finally, the aggregators in turn provide the desired balancing power in response to the price signals. However, there are some existing problems: the ISO could not know in advance how much balancing power the building aggregators are able to provide at most. Therefore, two outcomes are possible after running the DSO's bi-level optimization model: ① the value of the objective function in the optimal solution is very small and the stopping criterion in (7.34) is satisfied. It means the building aggregators are able to provide the enough balancing power; ② the value is very large which means that the buildings cannot provide the required balancing power.

$$\left(\varepsilon_{E,n}\left|\Delta P_{E,n}^{avi}-\Delta P_{E,n}^{req}\right|+\varepsilon_{H,n}\left|\Delta P_{H,n}^{avi}-\Delta P_{H,n}^{req}\right|\right)\leq\sigma \tag{7.33}$$

For the first situation, the clearing of the DRX market is achieved. The DSO just needs to send out the incentive prices to the aggregators after that. For the second situation which is much more likely to occur, the clearing of the DRX should be continued. The target values ($\Delta P_{E,n}^{req}$, $\Delta P_{H,n}^{req}$) should be adjusted since the building aggregators cannot provide so much balancing power and the bi-level optimization model cannot get the converged result. The new target values should satisfy two requirements: they should be achievable; the deviations from the original target values should be as small as possible, since the original target values realize the optimization of the DR-IHED model. Therefore, this chapter develops an optimum feasible region method to obtain the new target values, which is illustrated in Fig. 7.3.

After running the DR embedded IHED model, the required balancing power for different time periods can be obtained, which is illustrated in the upper half of Fig. 7.3. The DSO runs the optimization model to make the actual balancing power $\Delta P_{E,n}^{avi}$, $\Delta P_{H,n}^{avi}$ in line with the desired balancing power as possible as it can be done. After running the DSO's optimization model shown in (7.29)–(7.32), the optimum balancing power results can be obtained. Such optimum results, represented as $\left(\Delta P_{E,n}^{avi}(\mathrm{t1}),\cdots\Delta P_{E,n}^{avi}(\mathrm{t4});\Delta P_{H,n}^{avi}(\mathrm{t1}),\cdots\Delta P_{H,n}^{avi}(\mathrm{t4})\right)$, are shown in the lower half of Fig. 7.3. Using the optimum DR results as the vertices, the optimum feasible regions of the available electric balancing power heat balancing power can be obtained, which are given by the green area and red area in Fig. 7.3, respectively.

The feasible regions can be expressed as a set of constraints:

$$f_n(\Delta P_{E,n}^{avi}(\mathrm{t1}),\Delta P_{H,n}^{avi}(\mathrm{t1}),\cdots\Delta P_{E,n}^{avi}(NT_H),\Delta P_{H,n}^{avi}(NT_H))\leq 0 \tag{7.34}$$

Obviously, the buildings' comfort level can be guaranteed within the feasible region. Moreover, the minimum deviation between the required balancing power and the actual balancing power can be achieved since the results are generated by the

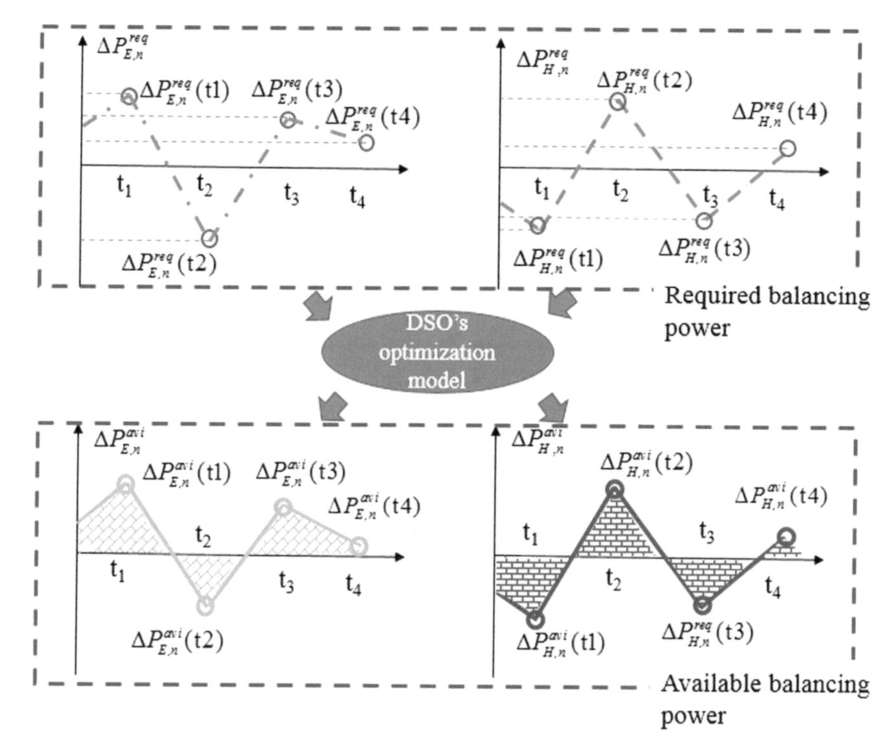

Fig. 7.3 Illustration of the feasible region method

bi-level optimization model. Therefore, the DSO can submit the feasible region of the available balancing power to the ISO. Then, the ISO runs the IHED model again considering the feasible region of the available balancing power to renew the target values. Since the new target values are within the feasible region, it can be guaranteed that the DSO is able to stimulate the aggregators to provide so much balancing power. Running the bi-level optimization model again, DR schedules corresponding to the new target values can be determined. Therefore, the converged results of the bi-level optimization problem are achieved and the stopping criterion would be satisfied.

The overall process of the market clearing is illustrated in Fig. 7.4. As shown in Fig. 7.4, the ISO firstly determines how much balancing power is required from the DRX market by utilizing the IHED model as shown in (7.16)–(7.27). Then, the desired DR power at each bus $\Delta P^{req}_{E,n}$, $\Delta P^{req}_{H,n}$ is sent out to the DSOs. The DSOs run the proposed bi-level optimization problem shown in (7.29)–(7.32) to make the real DR power $\Delta P^{avi}_{E,n}$, $\Delta P^{avi}_{H,n}$ in line with the desired DR power as possible as it can be done. Once the stopping criterion in (7.34) is satisfied, the DRX clearing is successful and will be terminated. Otherwise, the optimum feasible region of the available balancing power is obtained and submitted to the ISO. The ISO runs the DR-IHED model again with the consideration of the feasible regions. Finally, the new generation schedules and DR schedules are obtained.

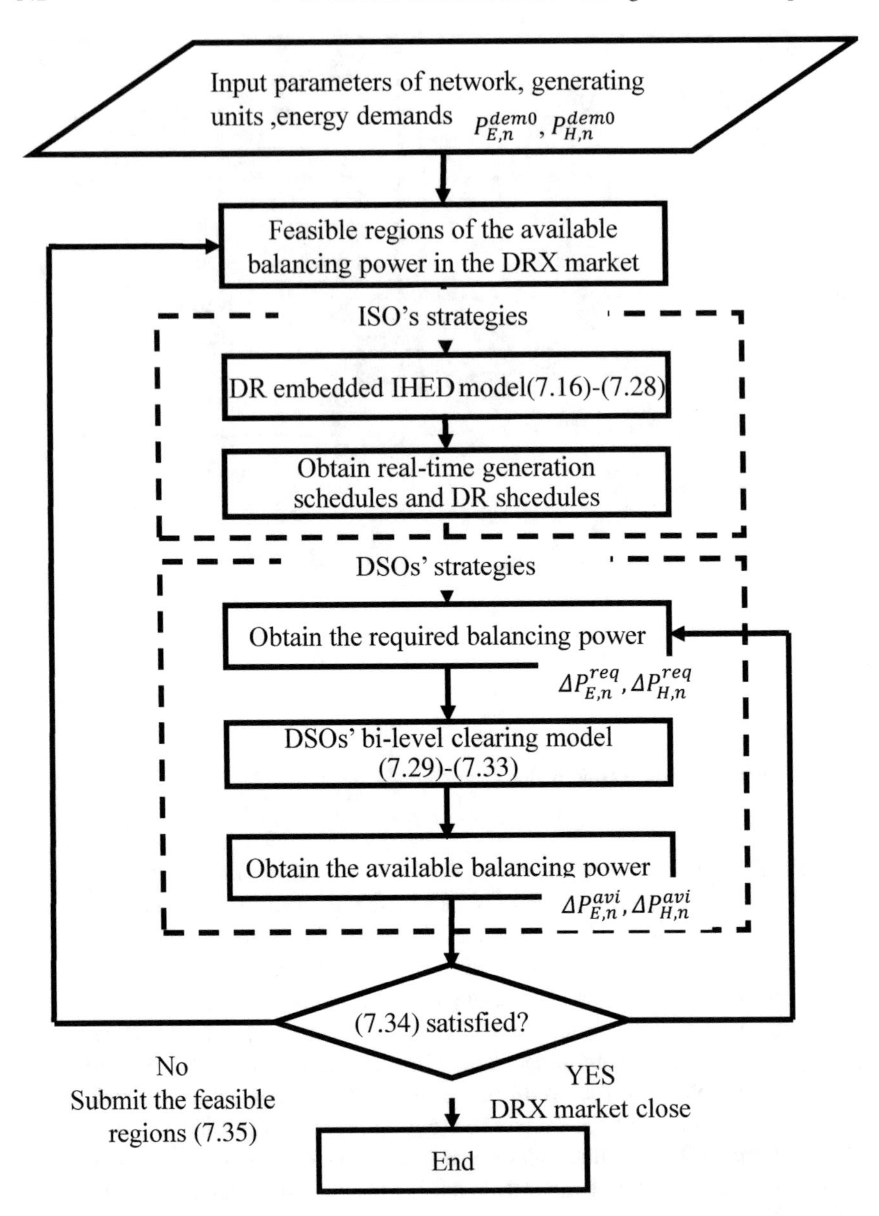

Fig. 7.4 Process of the market clearing

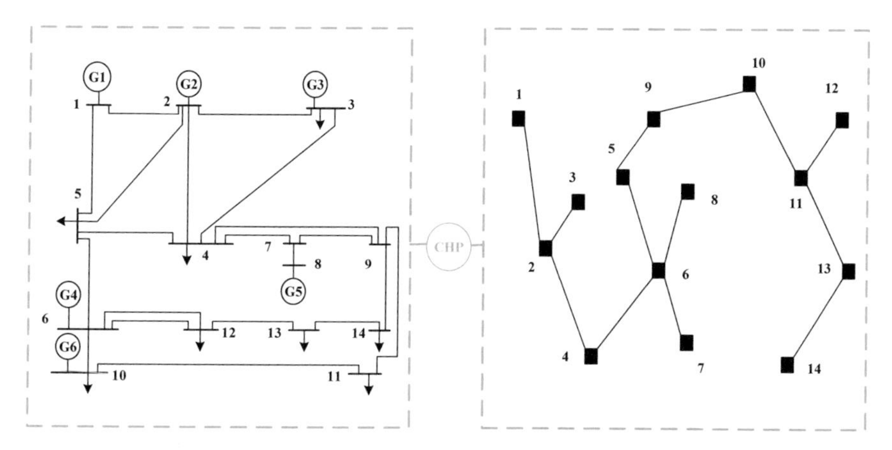

Fig. 7.5 Topology diagram of the integrated energy system

Table 7.1 Parameters of heating networks

	Value		Value
C_w	4200 (J/(kg°C))	$\tau_{r,i}$	60 (°C)
l	300 (m)	A	120 (m²)
$\tau_{s,i}$	85 (°C)	H	2.5 (m)

7.4 Case Studies

7.4.1 Test System and Parameters

A test system is introduced to illustrate the technique proposed in this chapter. The system is developed from the 14-bus electric power system [25] and a 14-node district heating system [26], with two CHP units (G1, G2), three coal-fired power generating units (G3–G5) and one wind farm (G6). The topological diagram of the energy system is shown in Fig. 7.5.

The profiles of the electricity demand (non-heating), wind power potential, and ambient temperature are shown in Fig. 7.6. The electricity demand profile is derived from [27]. The wind power profile is derived based on historic data of a wind farm located in northeast China.

The maximum aggregated electricity demand in each bus can be found in [25] and it is assumed that the heat demand has the same maximum value.

The parameters of the DH networks and buildings are shown in Table 7.1.

The heat transfer coefficients are:

$$u_{a-h} = 7.69A_s$$

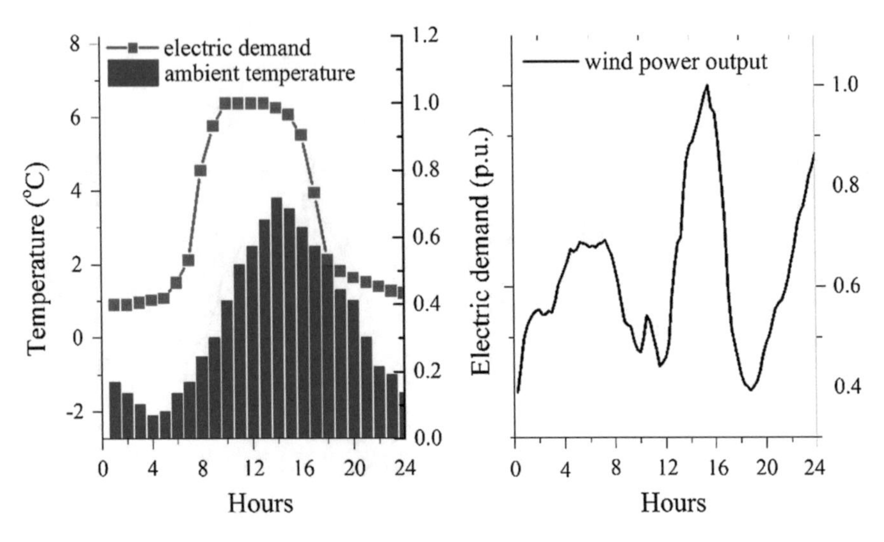

Fig. 7.6 Electricity and heat demand as well as wind power output

$$u_{a-o} = 0.17V_s$$

$$u_{h-o} = 7.69 \cdot A_s \cdot (69.05 + 1.07A) \big/ 7.69 \cdot A_s - (69.05 + 1.07A)$$

where A_s and V_s are the surface area and volume of the building, respectively.

Moreover, the rated electric power demand is set correspondingly as 4.8 kW [28]. The threshold of the indoor temperature is set as 18°C–24°C [29]. The simulation tool is based on a primal-dual interior point solver called MATLAB Interior Point Solver (MIPS). Moreover, we run the simulation on a PC with Intel 2.4 GHz 2-core processor (4 MB L3 cache), 8 GB memory.

7.4.2 Comparison Between the Proposed Comprehensive DR Strategies and the Load Shifting Strategy

In order to illustrate the benefits of the proposed technique, three scenarios (S1–S3) are modeled: S1 is the no-DR scenario; In Scenario S2, only load shifting strategy is utilized and buildings are expected to only provide the electric balancing power in scenario S2; the proposed comprehensive DR strategy is applied in scenario S3, where buildings are able to adjust their electric power demand and heat power demand through the load shifting and substitution strategies.

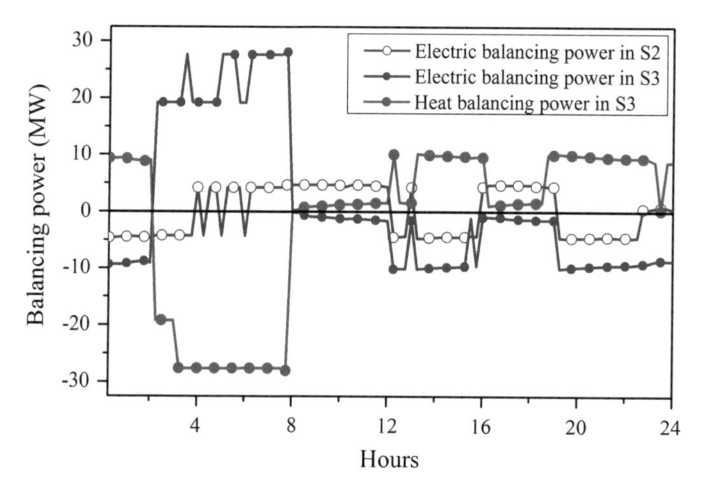

Fig. 7.7 Balancing power in different scenarios

(a) Balancing Power

In scenario S2, the buildings are able to adjust their electric power demand so as to provide the electric balancing power taking advantage of the thermal inertia of the buildings.

The balancing power provided by such DR behaviors in scenario S2 is shown in empty-circle line Fig. 7.7. As shown in Fig. 7.7, the peak capacity of the balancing power in scenario S2 is about 15 MW (absolute value).

The balancing power provided by the buildings in scenario S3 is shown as the solid-circle lines in Fig. 7.7. As we can see in the Fig. 7.7, more balancing power, including heat balancing power, can be provided in scenario S3. In Scenario S3, the peak capacity of the balancing power is about 30 (absolute value). It leads to the conclusion that the proposed comprehensive DR strategy is able to better exploit the demand flexibility of the buildings.

(b) Energy Price

A major benefit of DR is that it contributes to reducing the energy price volatility. The energy prices here refer to the real-time energy prices derived from the DR-IHED model. The energy prices from different scenarios over the 24-h horizon are depicted in Fig. 7.8. The energy prices without the consideration of DR (in S1) are shown as the empty-square lines in the Fig. 7.8.

As we can see in Fig. 7.8, the energy prices without DR fluctuate widely. The electricity prices fluctuate between 0$/MWh and 101$/MWh while the heat prices fluctuate between 20$/MWh and 102$/MWh. It is the fluctuation of wind power that makes the energy prices vary significantly, which may lead to a series of issues, e.g. the profit of CHP units can be remarkably reduced in periods of low electricity price resulting from large wind power production [30].

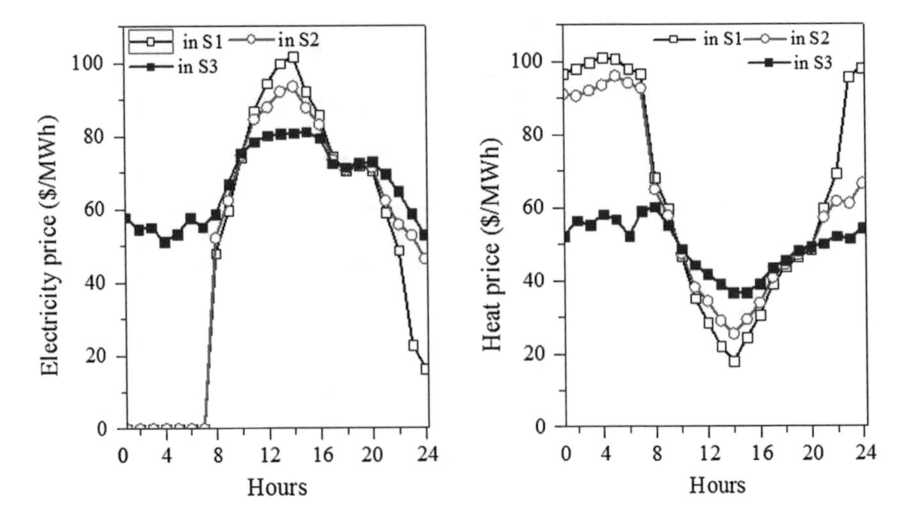

Fig. 7.8 Energy prices in different scenarios

Compared with the energy prices in the no-DR scenario, it can be seen that DR contributes to reducing the volatility of energy prices. In scenario S2 where only the electric balancing power is considered, the fluctuation ranges of electricity prices and heat prices are narrowed down to 0–90$/MWh and 25–92$/MWh, respectively. Moreover, the electricity price fluctuation range is further narrowed down to 45–80$/MWh while the heat price fluctuation range is further narrowed down to 38–61$/MWh in scenario S3. It can be concluded that the proposed method is better in mitigating the price volatility.

(c) Wind Power Integration

The other benefit of the DR is that it mitigates the wind power curtailment. The simulation results in term of wind power accommodation are depicted in Fig. 7.9.

It can be pointed out that a serious wind power curtailment occurs during off-peak hours in the no-DR scenario. To meet the customers' pre-determined constant heat demand, CHP units have to remain on and generate a certain amount of electricity, which occupies part of the proportion of wind power generation. When the DR is considered, the wind power curtailment can be mitigated. As shown in Fig. 7.9, the wind power can be utilized more effectively in scenario S2. The increased wind power integration in scenario S2 compared with scenario S1 is represented as the area "wind power increase 2". In scenario S3, the wind power curtailment can be mitigated more remarkably. The increased wind power integration in scenario S3 compared with scenario S2 is represented as the area "wind power increase 1". In the proposed method, substitution plays a crucial role in exploiting the demand flexibility of buildings during the off-peak hours. In the DRX market, the DSOs stimulate the customer aggregator to switch a part of their heating demand from DH-heat to the electric heating so as to increase their electric demand and decrease

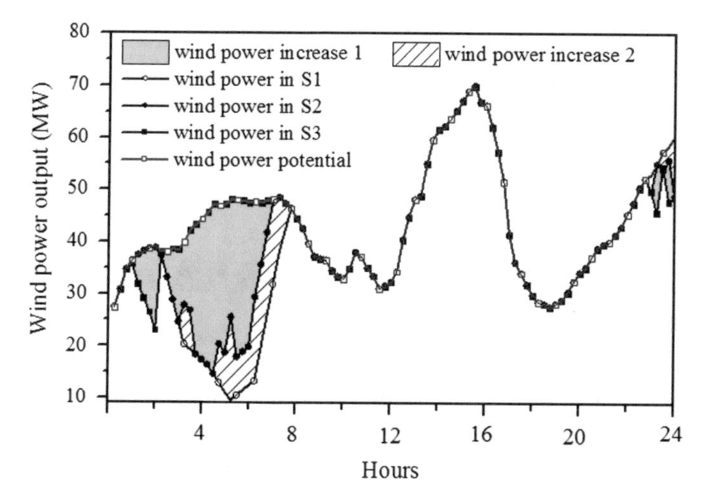

Fig. 7.9 Wind power output in different scenarios

Table 7.2 Social welfare ($)

Scenarios	S1	S2	S3
10% DR	173 124	174 123	176 898
20% DR		175 104	180 672
30% DR		175 852	183 417

their heat demand by setting a relatively low electricity price and a relatively higher heat price. The substitution-DR leads to a decrease in the electricity production from CHP units and an increase in the aggregated electricity demand. Thereby, the wind power generation can be effectively utilized.

(d) Social Welfare

Table 7.2 provides a numerical comparison of social welfare for the different scenarios. Moreover, the social welfare in S2 and S3 with different DR rates are also presented.

It can be seen from Table 7.2 that a higher social welfare can be achieved when the DR is considered in scenario S3. In scenario S2, the social welfare increases from 173 124$ to 175 852$ when 30% of the buildings are involved in the DR programs. In other word, the social welfare increases by 1.6% if only load-shifting DR is considered. In scenario S3 where the proposed method is applied, the social welfare further increases to 183 417$, which is 5.9% higher than that in the no DR scenario and 4.3% higher than that in the scenario S2. The reasons for the higher social welfare in scenario S3 include: the energy system can be operated with a higher flexibility and effectively since more balancing power can be provided; more wind power is utilized in scenario S3 which contributes to reduce the generation cost.

In summary, the first comparison demonstrates that compared with the load shifting strategy, the proposed comprehensive DR strategy is able to fully exploit the demand flexibility of buildings, which comes with several advantages, including ① the volatility of energy prices can be reduced; ② the wind power can be utilized more efficiently; ③ the social welfare can be increased more remarkably.

7.4.3 Comparison Between the Real-Time DRX Framework and the Day-Ahead DR Framework

The comparisons between the developed real-time DRX and the existing day-ahead DRX market are also made. For facilitating the comparison, two cases are developed: Case 1 represents the existing day-ahead DRX market and Case 2 represents the developed real-time DRX markets.

(1) **Energy Price**

The simulations results in the two cases are firstly compared in term of the energy price volatility as shown in Fig. 7.10. In Case 1 where the day-ahead DRX market is considered in each hourly interval, only the hourly balancing power can be provided by the buildings. Therefore, it is impossible to balance the very short-term, e.g. 15-min, power fluctuation from wind energy. In this case, the very short-term power fluctuation must be balanced by frequently adjusting the traditional generating units, which leads to the appearance of the price spikes. In Case 2, the DRX market is launched in each 15-min interval during the operational hour. Hence, a balancing power on a smaller time scale can be provided to absorb the very short-term power fluctuation from wind power. In this way, the price spikes are reduced.

(2) **Energy Generation Cost**

As introduced above, traditional generating units need to frequently adjust their power output to balance the very short-term power fluctuation from wind energy. Besides the appearance of price spikes, the other disadvantage coming with it is the higher energy generation costs. As shown in Fig. 7.11, the average generation cost in Case 1 is higher than that in Case 2. It can be calculated that the average generation cost in Case 1 is 29 840 \$/h and the average generation cost in Case 2 is 29 711 \$/h.

In summary, the second comparison demonstrates that the proposed real-time DRX market is able to balance the very short-term power fluctuation from wind energy through DR, which helps to reduce the price spike and reduce the generation costs.

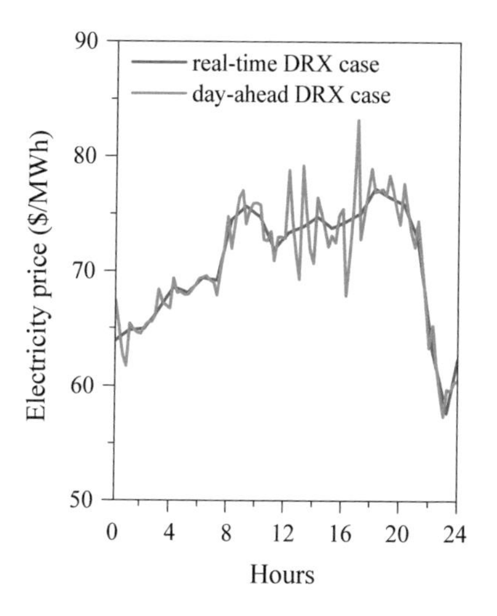

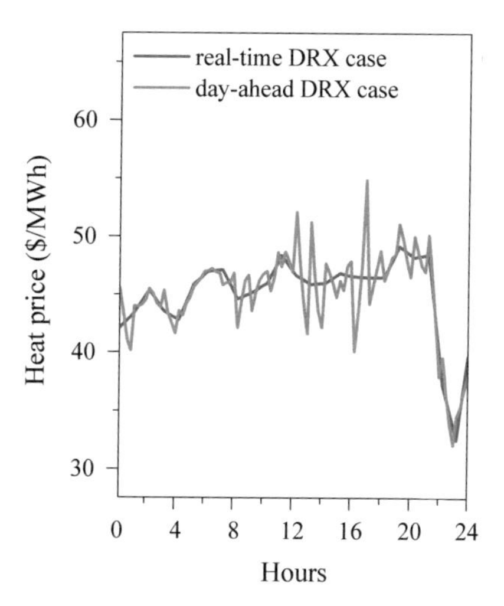

Fig. 7.10 Energy prices in the two cases

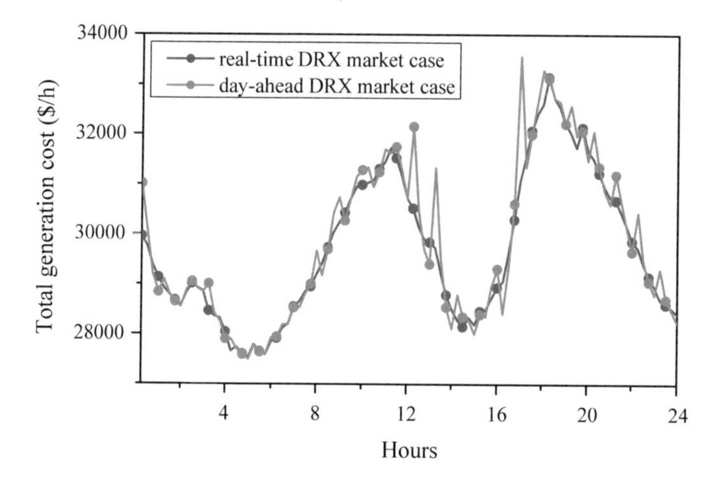

Fig. 7.11 Hourly total generation cost in the two cases

7.4.4 Comparisons Between the Proposed Clearing Method and the Iteration-Based Clearing Method

As introduced in previous sections, the iteration-based clearing methods have been proposed for clearing the DRX market. However, with an increasing number of the aggregators, it may lead to many iterations and longer computational time. Therefore, the method may not be able to meet the higher requirement for clearing speed in the real-time market. Given that, this chapter proposes a novel optimum feasible region method to achieve the fast clearing DRX market.

Table 7.3 Computational performance comparison

	Iteration times	CPU time(s)
IM: 20 aggregators	5	5.63
OFRM: 20 aggregators	2	2.65
IM: 30 aggregators	7	9.28
OFRM: 30 aggregators	2	2.65
IM: 50 aggregators	10	13.26
OFRM: 50 aggregators	2	2.65

Comparison between the iteration-based method (IM) and the proposed optimum feasible region method (OFRM) in term of computational performance is shown in Table 7.3. It can be seen that the proposed method requires less computation time and iterations. Moreover, the benefit is more obvious when the number of aggregators increases.

7.5 Conclusions

This chapter expands the DR concept to the HE-IES where the additional balancing power is required to integrate the variable wind power. The utilization of smart buildings to provide both the heat balancing power and electric balancing power is investigated. A comprehensive DR strategy is firstly proposed which combines the load shifting and energy substitution. Then, an aggregation method is proposed to aggregate multiple buildings' flexible energy demands. Moreover, a real-time DRX market is developed where the building aggregators are stimulated to adjust buildings' energy consumption behaviors and provide the required balancing power. Additionally, a novel optimum feasible region method is proposed to achieve the fast clearing of the DRX market to meet the higher requirement for clearing speed in the real-time market. The case studies demonstrate that: the proposed comprehensive DR strategy is able to fully exploit the demand flexibility of buildings; the proposed real-time DRX market is able to balance the very short-term fluctuation of wind power through DR, which helps to reduce the price spikes and the generation cost; the proposed optimum feasible method offers a better computational performance if compared with the traditional method.

References

1. G. Streckienė, V. Martinaitis, A.N. Andersen, J. Katz, Feasibility of CHP-plants with thermal stores in the German spot market. Appl. Energy **86**, 2308–2316 (2009)
2. P.F. Bach, Towards 50% wind electricity in Denmark: dilemmas and challenges. Eur. Phys. J. Plus **131**, 1–12 (2016)

3. X. Lu, M.B. McElroy, W. Peng, S. Liu, C.P. Nielsen, H. Wang, Challenges faced by China compared with the US in developing wind power. Nat. Energy **1**, 16061 (2016)
4. M.G. Nielsen, J.M. Morales, M. Zugno, T.E. Pedersen, H. Madsen, Economic valuation of heat pumps and electric boilers in the Danish energy system. Appl. Energy **167**, 189–200 (2015)
5. X. Lu, M.B. Mcelroy, W. Peng, S. Liu, C.P. Nielsen, H. Wang, *Challenges Faced by China Compared with the US in Developing Wind Power*, vol. 1 (2016), p. 16061
6. O. Erdinc, N.G. Paterakis, I.N. Pappi, A.G. Bakirtzis, J.P.S. Catalão, A new perspective for sizing of distributed generation and energy storage for smart households under demand response. Appl. Energy **143**, 26–37 (2015)
7. P.H. Shaikh, N.B.M. Nor, P. Nallagownden, I. Elamvazuthi, T. Ibrahim, A review on optimized control systems for building energy and comfort management of smart sustainable buildings. Renew. Sustain. Energy Rev. **34**, 409–429 (2014)
8. Y. Lin, P. Barooah, S. Meyn, T. Middelkoop, Experimental evaluation of frequency regulation from commercial building HVAC systems. IEEE Trans. Smart Grid **6**, 776–783 (2015)
9. C. Vivekananthan, Y. Mishra, G. Ledwich, F. Li, Demand response for residential appliances via customer reward scheme. IEEE Trans. Smart Grid **5**, 809–820 (2014)
10. E. Bilgin, M.C. Caramanis, I.C. Paschalidis, C.G. Cassandras, Provision of regulation service by smart buildings. IEEE Trans. Smart Grid **7**, 1683–1693 (2017)
11. D.T. Nguyen, M. Negnevitsky, M.D. Groot, Pool-based demand response exchange—concept and modeling. IEEE Trans. Power Syst. **26**, 1677–1685 (2011)
12. D.T. Nguyen, M. Negnevitsky, M.D. Groot, Walrasian market clearing for demand response exchange. IEEE Trans. Power Syst. **27**, 535–544 (2012)
13. H. Wu, M. Shahidehpour, A. Alabdulwahab, A. Abusorrah, Demand response exchange in the stochastic day-ahead scheduling with variable renewable generation. IEEE Trans. Sustain. Energy **6**, 516–525 (2015)
14. C. Shao, Y. Ding, P. Siano, Z. Lin, A framework for incorporating demand response of smart buildings into the integrated heat and electricity energy system. IEEE Trans. Ind. Electron. **66**(2):1465-1475 (2019).
15. M. Geidl, G. Andersson, Optimal power flow of multiple energy carriers. IEEE Trans. Power Syst. **22**, 145–155 (2007)
16. B. Daryanian, R.E. Bohn, R.D. Tabors, Optimal demand-side response to electricity spot prices for storage-type customers. IEEE Trans. Power Syst. **4**, 897–903 (1989)
17. P. Wang, J.Y. Huang, Y. Ding, P. Loh, L. Goel, Demand side load management of smart grids using intelligent trading/metering/billing system, in *IEEE PES General Meeting* (2010), pp. 1–6
18. R.C. Sonderegger, *Dynamic Models of House Heating Based on Equivalent Thermal Parameters* (1978)
19. X.S. Jiang, Z.X. Jing, Y.Z. Li, Q.H. Wu, W.H. Tang, Modelling and operation optimization of an integrated energy based direct district water-heating system. Energy **64**, 375–388 (2013)
20. Y. Ding, S. Pineda, P. Nyeng, J. Ostergaard, E.M. Larsen, Q. Wu, Real-time market concept architecture for EcoGrid EU—A prototype for European smart grids. IEEE Trans. Smart Grid **4**, 2006–2016 (2013)
21. Z. Li, W. Wu, J. Wang, B. Zhang, T. Zheng, Transmission-constrained unit commitment considering combined electricity and district heating networks. IEEE Trans. Sustain. Energy **7**, 480–492 (2016)
22. B. Awad, M. Chaudry, J. Wu, N. Jenkins, Integrated optimal power flow for electric power and heat in a MicroGrid, in *International Conference and Exhibition on Electricity Distribution* (2009), pp. 1–4
23. T. Orfanogianni, G. Gross, A general formulation for LMP evaluation. IEEE Trans. Power Syst. **22**, 1163–1173 (2007)
24. R. Jabr, A.H. Coonick, B.J. Cory, A primal-dual interior point method for optimal power flow dispatching. IEEE Power Eng. Rev. **22**, 55 (2002)
25. C. Wang, M.H. Nehrir, Analytical approaches for optimal placement of distributed generation sources in power systems. IEEE Trans. Power Syst. **19**, 2068–2076 (2004)

26. A. Shabanpour-Haghighi, A.R. Seifi, Simultaneous integrated optimal energy flow of electricity, gas, and heat. Energy Convers. Manag. **101**, 579–591 (2015)
27. L. Pedersen, *Method for Load Modelling of Heat and Electricity Demand* (2006)
28. H. Hui, Y. Ding, W. Liu, Y. Lin, Y. Song, Operating reserve evaluation of aggregated air conditioners. Appl. Energy **196**, 218–228 (2017)
29. Z. Pan, Q. Guo, H. Sun, Feasible region method based integrated heat and electricity dispatch considering building thermal inertia. Appl. Energy (2016)
30. T. Jónsson, P. Pinson, H. Madsen, On the market impact of wind energy forecasts. Energy Econ. **32**, 313–320 (2010)

Chapter 8
Economical Evaluation of the Flexible Resources for Providing the Operational Flexibility in the Power System

8.1 Introduction

There is a consensus that the intermittency and uncertainty of wind power have been the major barriers for large scale wind power integration. To deal with the uncertainty of wind power, many methods have been developed to improve wind power forecasting accuracy [1, 2]. Moreover, many studies have been conducted on how to improve the system flexibility so as to deal with the wind power intermittency, which usually combine the wind power with other flexible resources [3–5]. The previous chapters have fully analyzed the potential, framework and advantages of using demand-side resources to provide the flexibility which is necessary for integrating the wind power. Such measures can be categorized into demand-side management (DSM). Besides DSM, there are also two kinds of flexibility options for integrating the fluctuating wind power, including using the operating reserves from the conventional generation, and using the flexibility provided by the energy storage [6]. This paper evaluates the economy of the three different flexibility resources to find the advantages/disadvantages of different resources and to provide a guidance for investment in these flexible resources.

Traditionally, the flexibility from conventional thermal power generation for providing operating reserves is the most important option for integrating fluctuating wind power. For example, the flexibility for integrating the fluctuating wind power are typically provided by conventional generating units in China, such as coal-fired power generating units [7]. However, utilizing the conventional generating units to integrate the variable wind energy causes additional costs. Examples include the short-term balancing services, provision of firm reserve capacity, and more cycling and ramping of conventional plants for integrating the wind power [8]. Integrating the fluctuating wind power, the conventional generating units should work at part-load and change their output frequently to cope with the variability and uncertainties associated with wind energy. Consequently, the operation of generating units is varying and low load levels results in low energy efficiency, higher fuel consumption and the additional cost. The additional cost arising from

Y. Ding et al., *Integration of Air Conditioning and Heating into Modern Power Systems*,
https://doi.org/10.1007/978-981-13-6420-4_8

the intermittency of wind and the subsequent causation of 'balancing plants' for system security is widely observed [9]. Moreover, the increased fuel consumptions come with the additional carbon emissions. In other word, the effect of developing wind power in decarbonizing the power systems is partially offset by the additional carbon emissions due to providing required flexibility [6]. Fortunately, with the development of smart grid technologies, energy storage and DSM may be able to compete with the flexibility provided by the conventional generation. Actually, energy storage and DSM could be the preferred options since they avoid additional energy consumption and emissions. Recent advances in electric energy storage technologies provide an opportunity for using energy storage to address the wind energy intermittency. There are already some investigations on the use of energy storage applied to wind turbines for buffering the variability of the output. It is convinced that the energy storage system (ESS) is able to reduce the variability and uncertainty of short-term wind power [10], and lift the capacity credit of wind power [11], and increase the profit of the wind power [12]. Meanwhile, DSM can be another source of this required flexibility. Utilization of smart grid technologies in power systems creates opportunities to more efficiently balance supply and demand [5, 13]. It has been demonstrated that traditionally passive loads may become a resource that can mitigate the consequences of wind's variability [14]. Other studies have found that utilizing the DSM is able to compensate the wind power forecasting uncertainty as well as reduce the total operational cost and air pollutant emissions [15].

Providing the flexibility and balancing power to integrate the variable wind power results in additional cost, no matter which kind of flexible resources is used. There exist some studies analyzing and evaluating that cost which is referred to as the "balancing cost" or "integration cost" [8, 16–19]. The balancing cost is the cost of the flexible resources for integrating the wind power, expressed in dollars per megawatt-hour of wind power generation ($/MWhw). Moreover, most studies analyzed the balancing cost of wind power if the flexibility is provided by the existing generation resources. To the best knowledge of the authors, few studies have been conducted on evaluating the balancing cost when utilizing different flexible resources so as to find the most cost-effective way to integrate the variable wind power. Since the increasing penetration of wind power is redefining the requirement for flexibility, it is necessary to provide enough operational flexibility in the most economic manner. Therefore, this chapter evaluates the balancing cost when utilizing different flexibility options in order to find which the best option is from a cost perspective. In this way, guidance for investment in these flexible resources can be provided in this chapter for guaranteeing a pre-determined wind power development plan.

The research idea and contributions of this chapter can be summarized as Fig. 8.1. Firstly, a multi-state wind power model is developed to represent its variability and uncertainty. Then, the key indicators are proposed respectively for the different flexible resources to measure the balancing cost. Finally, the optimization models are developed to evaluate the indicators to find out the balancing cost when utilizing different flexible resources.

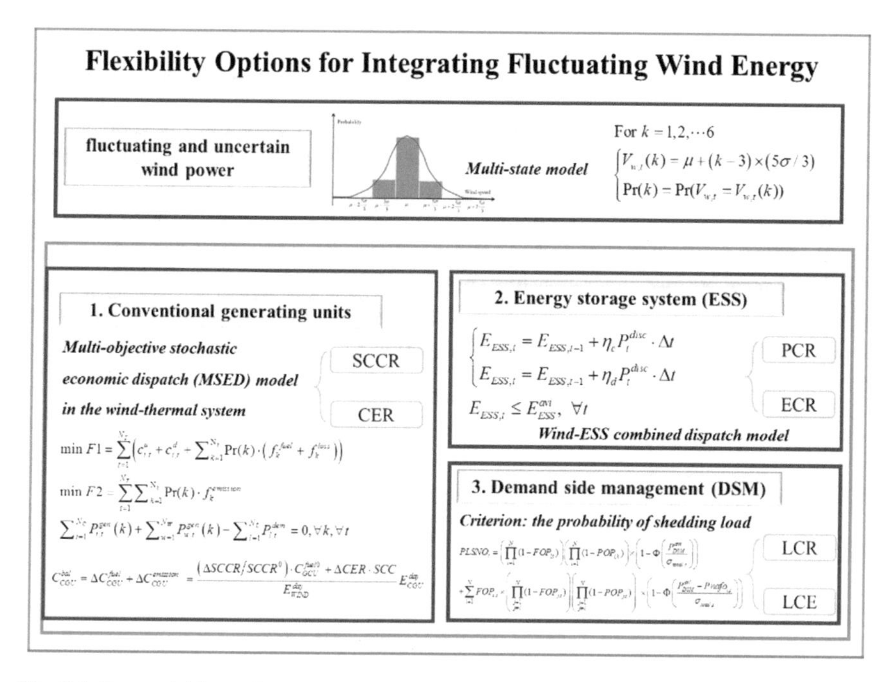

Fig. 8.1 Research idea and contributions of the study

(a) A multi-objective stochastic economic dispatch (MSED) is proposed to minimize the system operation cost and carbon emissions. Based on the MSED, the impact of the variable wind power on the conventional generation's energy efficiency of the can be assessed by comparing the with-wind scenario and without-wind scenario. Such negative impact is quantified using the system coal consumption rate (SCCR) index. Moreover, the SCCR can be converted to the carbon emission rate (CER). Consequently, the balancing cost when utilizing the conventional generating units can be measured by the additional fuel consumption and carbon emissions, which provides a benchmark and allows comparison with other resources. Moreover, the balancing cost also embodies the negative impact of the wind power's intermittency on the carbon emission reduction, which is not considered in other studies.

(b) An optimization model is developed to determine the power capacity requirement (PCR) and the energy capacity requirement (ECR) for integrating wind power using the ESS. The objective is to minimize the ESS's cost, including the capital cost and wear-out cost. Based on those parameters, the balancing cost of using ESS to integrate the wind power is also evaluated and compared with the benchmark cost.

(c) A reliability-constrained wind-DSM combined dispatch model is developed. Based on the model, the load power capacity requirement (LCR) and expected load-curtailment energy (LCE) of DSM programs for integrating wind power are also determined. Based on those parameters, the balancing cost of using DSM to integrate the wind power is evaluated and compared with the benchmark cost.

A case study is conducted to find the characteristics of using different flexible resources to integrate wind power. Conclusions are given in the Case Study section, which could provide guidance on the investment and future studies. It should be noted that this chapter takes China as an example for economically evaluating the key flexibility options for satisfying the wind power development target because wind power integration is especially significant in China [20]. However, the methods and models can be applied worldwide. This chapter includes research related to the economical flexibility options for integrating fluctuating wind energy in power systems by [21].

8.2 Methods to Calculate the Balancing Costs When Utilizing Different Flexible Resources

This section describes the mathematical models, based on which the balancing costs of utilizing different flexible resources can be determined.

8.2.1 *Mathematic Model for Evaluating the Balancing Cost of Utilizing Coal-Fired Generating Units*

(1) **Impact of Wind Power on the Operation of the Coal-Fired Generating Units**

Traditionally, the generation portfolio was designed to provide sufficient flexibility to cope with the variability and the forecast error of electric demand. In this case, the power balance can be formulated as:

$$\sum\nolimits_{i=1}^{N_C} P_{i,t}^{gen} - \sum\nolimits_{l=1}^{N_L} P_{l,t}^{dem} = 0,\ \forall t \tag{8.1}$$

where $P_{i,t}^{gen}$ denotes the power output of generating unit i at time t, and $P_{l,t}^{dem}$ denotes the electric demand of load l at time t.

As the generation capacity from wind power increases, the system also needs to be able to cope with the variability and uncertainties associated with the wind power [22, 23]. In China's coal-dominated power system, the flexibility requirements are usually met through operating reserves provided by coal-consumption generating units. The power balance in this case can be formulated as:

$$\sum\nolimits_{c=1}^{N_C} P_{i,t}^{gen} - \left(\sum\nolimits_{l=1}^{N_L} P_{l,t}^{dem} - \sum\nolimits_{w=1}^{N_W} P_{w,t}^{gen}\right) = 0,\ \forall t \tag{8.2}$$

where $P_w^{gen}(t)$ denotes the power output of wind power unit w at time t, and $\sum_{l=1}^{N_L} P_{l,t}^{dem} - \sum_{w=1}^{N_W} P_{w,t}^{gen}$ is referred to as the "net load".

As shown in (8.2), the conventional generating units should work at part-load and change their output frequently with the variation in customer demand as well as with the variability and uncertainties associated with wind energy. Moreover, the operation of generating units at varying and low load levels results in increased fuel consumption and associated carbon emissions.

(2) **Multi-state Wind Power Generation Model**

This section develops a multi-state wind power generation model to describe the variability and uncertainty of wind power.

The generated power of a wind turbine varies with the wind speed at the wind farm site. The power output of a wind turbine can be determined from its power curve, which is a plot of output power against wind speed. Equation (8.3) is the mathematical expression for a typical power curve of a wind turbine [24].

$$P_{w,t}^{avi} = \begin{cases} 0, & 0 \le V_{w,t} \le V_{ci} \\ P_w^{rate}\left(\mathrm{A} + \mathrm{B} \times V_{w,t} + \mathrm{C} \times V_{w,t}^2\right), & V_{ci} \le V_{w,t} \le V_r \\ P_w^{rate}, & V_r \le V_{w,t} \le V_{co} \\ 0, & V_{co} \le V_{w,t} \end{cases} \tag{8.3}$$

As shown in Eq. (8.3), the wind turbine starts generating at the cut-in speed V_{ci}, and is shut down at the cut-out speed V_{co}. Rated power P_w^{rate} can be obtained when the wind speed is between the rated speed V_r and the cut-out speed V_{co}. Moreover, there is a nonlinear relationship between the power output and the wind speed when the wind speed lies within the cut-in speed V_{ci} and the rated speed V_r as shown in Eq. (8.3), where the constants A, B, and C are presented in [24].

As discussed above, the power output of the wind turbine can vary continuously and intermittently from zero to the rated value depending on the wind speed at the wind farm site. Wind turbines are therefore usually represented by multistate models in analytical methods [25]. In the multi-state wind generation model, the wind speed is represented by a large number of discrete speed states. The model can be simplified by reducing the number of states at the cost of accuracy. It has been proved that the 6-state common wind speed model can be used for reliability studies of power systems with reasonable accuracy [26]. Therefore, a simplified multistate power generation model for a wind power plant can be determined by combining the 6-state common wind speed model with the power curve as shown in Eq. (8.3). The 6-state wind speed model is shown in Fig. 8.2, in which the six wind speed states and the corresponding probabilities are given.

Based on the 6-state wind speed model, the wind speeds for of a wind power plant can be obtained from this model using (8.4)

For $k = 1, 2, \ldots 6$

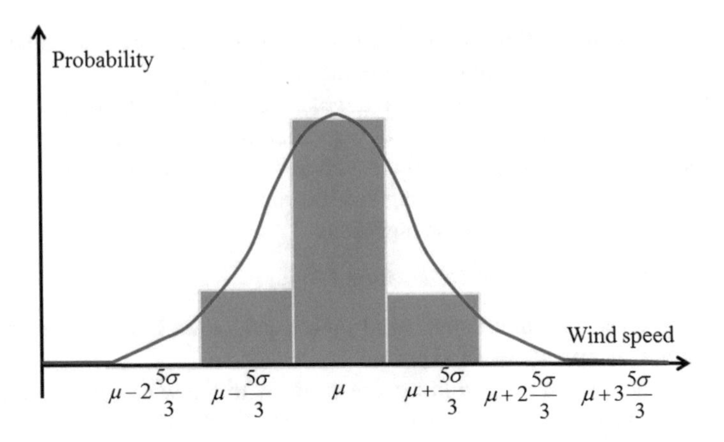

Fig. 8.2 Six-state wind speed model

$$\begin{cases} V_{w,t}(k) = \mu + (k-3) \times (5\sigma/3) \\ \Pr(k) = \Pr(V_{w,t} = V_{w,t}(k)) \end{cases} \tag{8.4}$$

where μ and σ are the expected and variance of the wind speed distribution. $\Pr(k)$ for the six states can be found in [26].

(3) **Optimal Dispatch of the Coal-Wind System**

As discussed above, a MSED is developed with different wind power penetration levels to quantify the incremental coal consumption of the generating units providing operating flexibility. The first objective function of the MSED is the overall system cost expressed in Eq. (8.5), including the start-up cost and turn-off cost of the generating units, the fuel consumption cost and the loss of load cost.

$$\min F1 = \sum_{t=1}^{N_T} \left(c_{i,t}^{u} + c_{i,t}^{d} + \sum\nolimits_{k=1}^{N_k} \Pr(k) \cdot \left(f_k^{fuel} + f_k^{loss} \right) \right) \tag{8.5}$$

where

$$f_k^{fuel} = \sum_{t=1}^{N_T} \left\{ \sum_{i=1}^{N_C} \left(a_i^{gen} + b_i^{gen} P_{i,t}^{gen}(k) + c_i^{gen} \left(P_{i,t}^{gen}(k) \right)^2 \right) + c_w^{cur} \left(P_{w,t}^{avi}(k) - P_{w,t}^{gen}(k) \right) \right\} \tag{8.6}$$

$$f_k^{loss} = \sum_{t=1}^{N_T} \sum\nolimits_{l=1}^{N_L} r_l^{loss} \left(P_{l,t}^{serve} - P_{l,t}^{dem} \right) \tag{8.7}$$

In (8.6), a_i^{gen}, b_i^{gen}, c_i^{gen} are the fuel consumption coefficients.

The second objective function is the expected carbon emission, which is denoted by Eq. (8.8)

$$\min F2 = \sum_{t=1}^{N_T} \sum_{k=1}^{N_k} \Pr(k) \cdot f_k^{emission} \tag{8.8}$$

where $f_k^{emission}$ is expressed as:

$$\sum_{t=1}^{N_T} \sum_{i=1}^{N_C} \left(a_i^{emission} + b_i^{emission} P_{i,t}^{gen}(k) + c_i^{emission} \left(P_{i,t}^{gen}(k) \right)^2 \right) \tag{8.9}$$

In (8.9), $a_i^{emission}$, $b_i^{emission}$, $c_i^{emission}$ are the carbon emission coefficients.

Moreover, by introducing the social cost of carbon (SCC), the second objective function can be also represented by a social cost. Therefore, the proposed MSED can be converted to a single-objective optimization model.

The power balancing constraints are expressed as:

$$\sum_{i=1}^{N_C} P_{i,t}^{gen}(k) + \sum_{w=1}^{N_W} P_{w,t}^{gen}(k) - \sum_{l=1}^{N_L} P_{l,t}^{dem} = 0, \forall k, \forall t \tag{8.10}$$

The ramp-rate limits of the generating units are expressed as:

$$\begin{aligned} P_{i,t}^{gen}(k_t) - P_{i,t-1}^{gen}(k_{t-1}) &\le R_i^u I_{i,t-1} + S_i^u \left(I_{i,t} - I_{i,t-1} \right) \forall t; k_t, k_{t-1} \in \{1, 2, k, \ldots N_k\} \\ P_{c,t-1}^{gen}(k_{t-1}) - P_{c,t}^{gen}(k_t) &\le R_i^d I_{c,t-1} + S_i^d \left(I_{i,t-1} - I_{i,t} \right) \forall t; k_t, k_{t-1} \in \{1, 2, k, \ldots N_k\} \end{aligned} \tag{8.11}$$

The generation output limits for the conventional generating units are expressed as:

$$I_{i,t} P_i^{\min} \le P_{i,t}^{gen}(k) \le I_{i,t} P_i^{\max}, \ \forall t, \forall k \tag{8.12}$$

The generation output limits for the wind power are expressed as:

$$0 \le P_{w,t}^{gen}(k) \le P_{w,t}^{avi}(k), \forall k \tag{8.13}$$

where $P_{w,t}^{avi}(k)$ is calculated based on the above multi-state wind power model:

$$P_{w,t}^{avi}(k) = \{ f_{V-P}(\mu + (k-3) \times (5\sigma/3)) \}, \forall k \tag{8.14}$$

The impact of wind power on the energy efficiency of coal-fired power generating units is quantified using the SCCR and CER indexes. The SCCR of with-wind and without-wind scenarios can be obtained based on the two-stage dispatch model, and then converted to the CER. Moreover, the additional carbon emissions can also be represented by a social cost. In this way, the total balancing cost of utilizing coal-fired generating units to integrate wind power can be expressed as:

$$C_{CGU}^{bal} = \Delta C_{CGU}^{fuel} + \Delta C_{CGU}^{emission} = \frac{(\Delta SCCR/SCCR^0) \cdot C_{GCU}^{fuel0} + \Delta CER \cdot SCC}{E_{WIND}^{day}} E_{CGU}^{day} \tag{8.15}$$

where $SCCR^0$ and C_{GCU}^{fuel0} refer to the SCCR and fuel cost of the coal-fired generating units without wind power integration. ΔCER and SCC denote the increase in carbon emissions and the social cost of carbon, respectively.

In (8.15), E_{CGU}^{day} and E_{WIND}^{day} are the total power generation of the coal-fired generating units and the wind power plants during the day, respectively. E_{CGU}^{day} and E_{WIND}^{day} are calculated as:

$$E_{CGU}^{day} = \sum_{t=1}^{N_T} \sum_{i=1}^{N_C} P_{i,t}^{gen}, \ E_{CGU}^{day} = \sum_{t=1}^{N_T} \sum_{w=1}^{N_W} P_{w,t}^{gen} \tag{8.16}$$

8.2.2 Optimization Model for Sizing the ESS and Determining the Balancing Cost

In addition to using operating reserves from conventional generating units, wind power variability can be operationally mitigated using energy storage. In fact, energy storage appears to be an obvious option to deal with the variability of renewable sources and the unpredictability of their output [27, 28]. In multiple application areas around the world, energy storage systems (ESSs) have been deployed to aid the integration of renewable energies, especially wind power [29]. Large-scale energy storage at the output of a wind farm can be used to mitigate the variability and uncertainty of wind power, reducing the need for coal-fired reserve generation and avoiding an increase in coal consumption. Moreover, certain energy storage technologies are already cost-competitive with certain conventional alternatives. Other energy storage technologies are also close to being cost-competitive in other applications, and the costs are expected to decline in the coming years. The projected costs of different storage technologies from the U.S. Energy Information Administration (EIA), Bloomberg New Energy Finance (BNEF) [30], AOE and financial advisory firm Lazard [31] are shown in Fig. 8.3.

The cost projections of different storage technologies used in different areas, including renewable energy generation integration (REN) and application in a transmission system (Trans), are derived from the financial advisory firm Lazard. The solid symbols and hollow symbols are the lower and upper bounds of the cost projections, respectively. The lines represent the cost projections come from the Energy Information Administration (EIA), Bloomberg New Energy Finance (BNEF) and Navigant.

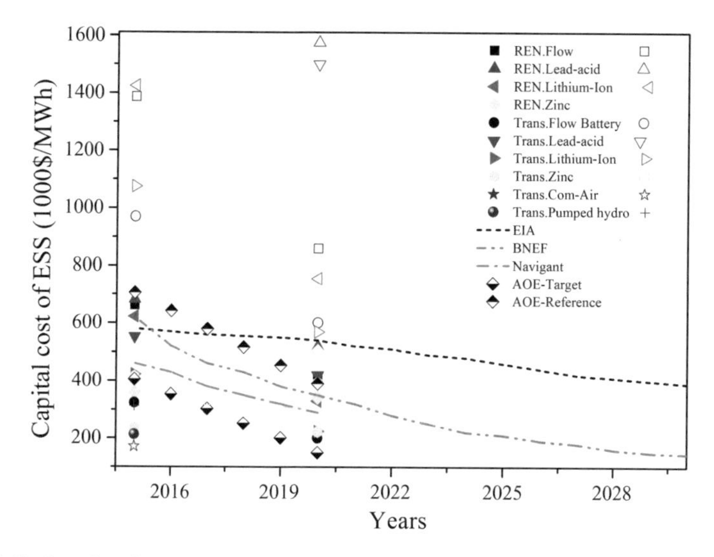

Fig. 8.3 Projected capital cost of different energy storage technologies

With the help of energy storage, the variation of wind power, can be smoothed out and the mismatch between the available renewable power and the load can be addressed [32]. Considering energy storage systems, the power balance can be formulated as:

$$\sum_{i=1}^{N_C} P_{i,t}^{gen} + \sum_{w=1}^{N_W} P_{w,t}^{gen} + P_t^{disc} - \sum_{l=1}^{N_L} P_{l,t}^{dem} = 0, \ \forall t \tag{8.17}$$

where P_t^{disc} denotes the discharging power of the ESS.

In this chapter, an optimization model is developed to determine PCR and ECR for using ESS to integrate wind power. The PCR is expressed in megawatts per megawatt of installed wind power. The ECR is expressed in, megawatt-hours per megawatt-hour of electricity production from wind power. The model to evaluate PCR (P_{ESS}^{avi}) and ECR (E_{ESS}^{avi}) is formulated as below.

The goal is to minimize the ESS's cost while making the wind-ESS combined output to meet an hour-ahead predicted power output [33].

$$\min C_{ESS}^{total} = \sum_{t=1}^{N_T} \left(c_t^{cap} + c_t^{wear} \right) \tag{8.18}$$

where c_t^{cap} is the capital cost divided into the period t, c_t^{wear} is the additional cost due to the rapid battery wear resulting from the deeper discharge in the wind-integration application [34].

c_t^{cap} and c_t^{wear} can be expressed as:

$$c_t^{cap} = \frac{p_E E_{ESS}^{avi} \cdot \Delta t}{\sum_{y=1}^{N_Y} \frac{365}{(1+r)^y}}; c_t^{wear} = \left| \int_{S_t}^{S_{t+1}} w(s)\mathrm{d}s \right| \tag{8.19}$$

In (8.19), p_E is the per-unit cost of the ESS, expressed in \$/MWh. w(s) denotes the wear-out density function, expressed as [35]:

$$w(s) = \frac{p_E}{2 \times E_{ESS}^{avi} \times \mu^2} \times \frac{b \times (1-s)^{b-1}}{a} \tag{8.20}$$

In (8.20), s is the state of charge (SOC) of the ESS, a and b are the specific coefficients, μ is the ESS's efficiency [36].

The power balance constraint is expressed in(8.17).

The charging and dis-charging characteristics of the ESS can be expressed as:

$$\begin{aligned} &\begin{cases} E_{ESS,t} = E_{ESS,t-1} + \eta_c P_t^{disc} \cdot \Delta t & P_t^{disc} \le 0, \text{ charge} \\ E_{ESS,t} = E_{ESS,t-1} + \eta_d P_t^{disc} \cdot \Delta t & P_t^{disc} > 0, \text{ discharge} \end{cases} \\ &\begin{cases} 0 \le P_t^{disc} \le P_{ESS}^{avi} \ \forall t, \text{discharge} \\ -P_{ESS}^{avi} \le P_t^{disc} \le 0 \quad \forall t, \text{charge} \end{cases} \\ &E_{ESS,t} \le E_{ESS}^{avi}, \ \forall t \end{aligned} \tag{8.21}$$

By running the above optimization model, the total ESS cost C_{ESS}^{total} can be determined. Then, the balancing cost of utilizing energy storage to integrate the wind power is the total cost of installing and operating an energy storage project divided by the wind power system over its life. The expressing of C_{ESS}^{bal} is given by

$$C_{ESS}^{bal} = C_{ESS}^{total} \bigg/ \sum_{y=1}^{N_Y} \frac{E_{WIND}^{day} \cdot 365}{(1+r)^y} \tag{8.22}$$

In (8.22), N_Y represents the expected lifetime of the ESS.

8.2.3 Optimization Model for Determining the Balancing Cost When Utilizing the DSM

With the development of DSM-enabled technologies, demand side resources hold an untapped potential for increasing system flexibility and aiding the integration of fluctuating wind power [37, 38]. Various types of DSM programs have been implemented in China; interruptible load is one example. Interruptible load represents a consumer load that, in accordance with contractual arrangements, can be interrupted at the time of annual peak load by the action of the consumer at the direct request of the system operator. While generators offer operational flexibility by providing the ability to increase their energy output, the load facilities, in contrast, provide the ability to reduce their energy consumption to offer a reserve. Wind power uncertainty can be managed at a lower cost through this type of DSM program to address wind forecast errors. For example, when the real-time wind power output is lower than the forecasted level, reduced energy consumption by a load facility addresses the demand and supply imbalances in the system. In this way, DSM provides the flexibility historically provided by coal-fired generating units. The power balance considering the DSM can be expressed as:

$$\sum_{i=1}^{N_C} P_{i,t}^{gen} + \sum_{w=1}^{N_W} P_{w,t}^{gen} + \sum_{l=1}^{N_L} P_{l,t}^{inter} - \sum_{l=1}^{N_L} P_{l,t}^{dem} = 0, \ \forall t \tag{8.23}$$

To utilize DSM as the reserve for integrating wind power, it is necessary to determine how much standby demand capacity is required. In addition, the expected load-curtailment energy must be evaluated to determine the energy capacity value of flexible demand, to provide guidance on setting the interruptible tariffs that compensate consumers for voluntary demand reductions. A quantitative model is developed to determine LCR and LCE of DSM programs for integrating wind power with different levels of penetration. The LCR (P_{DSM}^{avi}) is expressed in percentage of the required flexible load among the total electric demand. The LCE (E_{DSM}^{avi}) is expressed in, megawatt-hours per megawatt-hour of electricity production from wind power.

In the model, the DSM is introduced to mitigate the probability of shedding load (PLSNO) arising from the variability and uncertainty of electric demand and wind power. PLSNO comprises three components, as shown in (8.24) [39].

(a) The probability of not having any generator trip while having an un-forecasted wind and load variation greater than the system reserve level. This scenario corresponds to the first term in (8.24).
(b) The probability of having only one full generator trip and an un-forecasted wind and load variation greater that the system reserve level. This scenario corresponds to the second term in (8.24), which corresponds to the probability of having a wind and load variation greater than P_{DSM}^{avi} minus the power not available after the full outage of generator i, $Pnafo_{i,t}$.
(c) The probability of having only one partial generator trip and an unforecasted wind and load variation greater than P_{DSM}^{avi}. This scenario is similar to that in b) and corresponds to the third term in (8.24).

$$PLSNO_t = \left(\prod_{i=1}^{N}(1 - FOP_{i,t})\right)\left(\prod_{i=1}^{N}(1 - POP_{i,h})\right) \times \left(1 - \Phi\left(\frac{P_{DSM}^{avi}}{\sigma_{total,t}}\right)\right)$$
$$+ \sum_{i=1}^{N} FOP_{i,t} \times \left(\prod_{\substack{j=1 \\ j \neq i}}^{N}(1 - FOP_{j,t})\right)\left(\prod_{\substack{j=1 \\ j \neq i}}^{N}(1 - POP_{j,t})\right) \times \left(1 - \Phi\left(\frac{P_{DSM}^{avi} - Pnafo_{i,t}}{\sigma_{total,t}}\right)\right)$$

$$+\sum_{i=1}^{N} POP_{i,t} \times \left(\prod_{j=1}^{N}(1-FOP_{j,t})\right)\left(\prod_{\substack{j=1\\ j\neq i}}^{N}(1-POP_{j,t})\right) \times \left(1-\Phi\left(\frac{P_{ESS}^{avi}-Pnafo_{i,t}}{\sigma_{total,t}}\right)\right) \tag{8.24}$$

where $\Phi(x)$ denotes the normalized Gaussian distribution function of the system forecast error and $\sigma_{total,t}$ denotes the standard deviation of the total system forecast error. The load forecast error in time t can be modeled as a Gaussian stochastic fluctuating with a mean of zero and a standard deviation of $\sigma_{l,t}$. Since it is assumed that both the load and wind power forecast errors are uncorrelated Gaussian stochastic variables, the standard deviation of the total system forecast error $\sigma_{total,t}$ can be given by

$$\sigma_{total,t} = \sqrt{\sigma_{wind,t}^2 + \sigma_{load,t}^2} \tag{8.25}$$

The objective is to minimize the cost associated with the LCE while guaranteeing PLSNO within the acceptable ranges.

$$\min C_{DSM} = p_{DSM} E_{DSM}^{avi} \tag{8.26}$$

$$PLSNO_h \leq \varpi \tag{8.27}$$

The LCE is expressed as:

$$E_{DSM}^{avi} = \sum_{t=1}^{NT} \sum_{l=1}^{N_L} P_{l,t}^{inter} \tag{8.28}$$

After obtaining C_{DSM} based on the above model, the balancing cost of utilizing the DSM to integrate the wind power is expressed as:

$$C_{DSM}^{bal} = C_{DSM} \Big/ E_{WIND}^{day} \tag{8.29}$$

8.3 Simulation Results and Analysis

8.3.1 Parameters

There is no doubt that the simulation results, including the balancing costs, depend on the wind power patterns. The variations of wind occur on different time scales from seconds to seasons. Hence, the wind power data covering the entire year is necessary for obtaining the convincing results. In this chapter, the daily wind power output profiles are obtained from actual historic data from the wind farms located in North China. Moreover, to cover the necessary information that describes the variability of

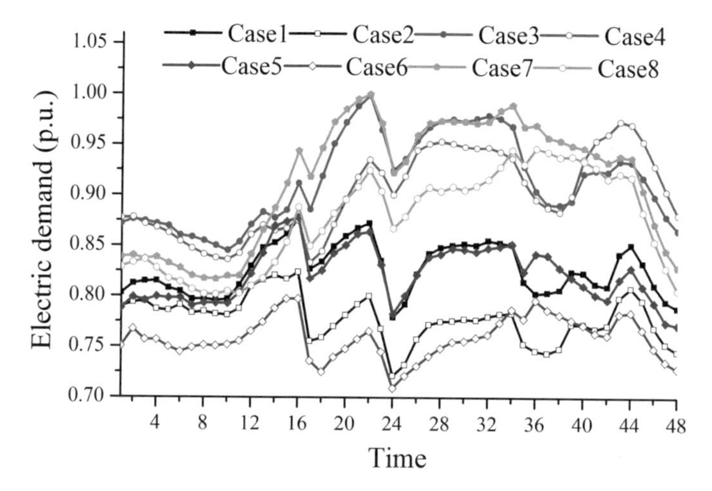

Fig. 8.4 Electric demand profiles

wind power, 337 sets of daily wind power profiles are selected and used in this study. The standard electric demand profiles derived from the historical data are shown in Fig. 8.4.

In Fig. 8.4, Case 1 to Case 8 cover four seasons: spring (Case 1 and Case 2), summer (Case 3 and Case 4), autumn (Case 5 and Case 6) and winter (Case 7 and Case 8). The weekday and weekend scenarios are also separated.

In this chapter, the coal-fired generating units are classified into three types according to their capacities: 300 MW units, 600 MW units and 1000 units. Based on the composition of the China's coal-fired generating units in 2015, the shares of the three types generating units are set as 44.8, 46.1 and 10.1%, respectively [40]. The coal consumption curves of the generating units with 300, 600 and 1000 MW capacities are shown in Fig. 8.5 [33].

8.3.2 Simulation Results

(1) Balancing Cost of Utilizing Coal-Fired Generating Units to Integrate Wind Power

In this section, the optimal system dispatch model is simulated with different wind power penetration levels to quantify the incremental coal consumption of the generating units providing flexibility for integrating wind power. Wind energy already generates 4% of China's electricity and will expand its share to 6% before 2020. Moreover, on 30 June 2015, China submitted its "Intended Nationally Determined Contribution" (INDC), including the target to peak carbon emissions by 2030 at the latest, and increase the share of non-fossil energy carriers of the total primary

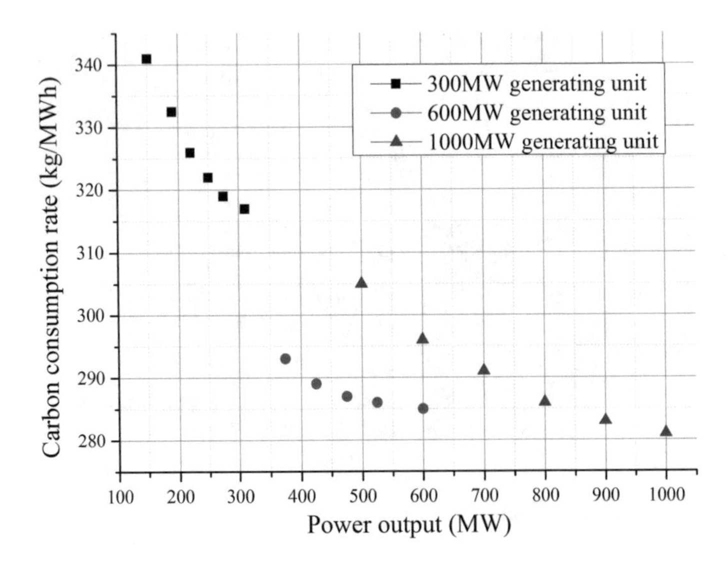

Fig. 8.5 Coal consumption curves of generating units with different capacities

energy supply to approximately 20% by that time [33]. Therefore, five scenarios are developed with different wind power penetration levels: no wind (reference case), 4% wind, 6% wind (2020 target), 10% wind (2030 target) and 20% wind (high wind power penetration case). Moreover, both the incremental fuel consumption and carbon emissions are transformed into costs, which are further defined as the balancing cost of wind power integration.

The impact of wind power integration on the energy efficiency of coal-fired power generating units is quantified using the SCCR index. The SCCRs for the "4% wind" scenario, the "6% wind" scenario, the "10% wind" scenario and the "20% wind" scenario are compared with the reference scenario where there is no integration of wind power. Moreover, the SCCR can be converted to the CER, which denotes the ton CO_2 per megawatt-hour of electricity generation.

The CERs of the different scenarios are shown in Fig. 8.6. The integration of wind power increases the carbon emission rate significantly. In the 4% wind and 6% wind scenarios, the carbon emission rates increase from 0.856 tCO_2/MWh to 0.858 tCO_2/MWh and 0.859 tCO_2/MWh, respectively. The effect is more obvious when the share of wind power integration is expanded between 10 and 20%. In 10% wind scenario, the SCCR increases to 0.861 tCO_2/MWh, while the carbon emission rate further increases to 0.869 tCO_2/MWh for the 20% wind scenario.

The increased coal consumption and its associated carbon emissions should not be ignored due to the predictable growth of wind power integration. In the 6% wind scenario, which can be achieved by 2020, the additional carbon emissions are estimated to be 15.56 million tons per year. When the wind power penetration level reaches 10%, which can be expected by 2030, the additional carbon emissions will further increase to 26.87 million tons per year.

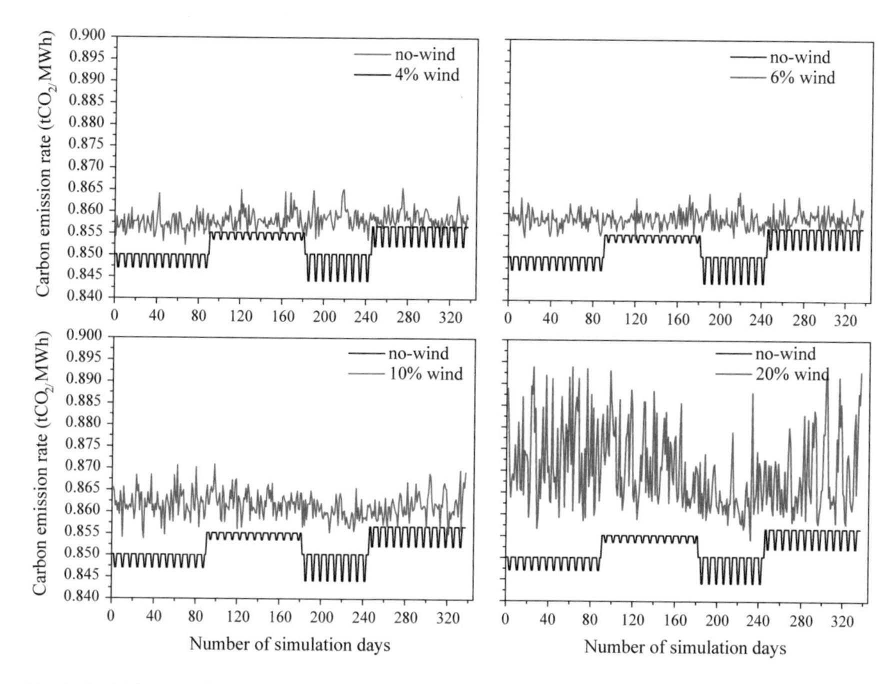

Fig. 8.6 CERs for different scenarios, where 4, 6, 10 and 20% are the penetration levels of wind power

The influences of integrating fluctuating wind power on the coal consumption rate, carbon emissions and balancing cost are summarized as shown in Fig. 8.7.

Obviously, SCCR, CER and the balancing cost are influenced by and are positively correlated with the wind power penetration level. It can be concluded that on average, a 1-MWh production of wind power will result in an increase of coal consumption of 13.68 to 17.64 kg depending on the wind power penetration level. Moreover, based on the simulation results, it can be concluded that the balancing cost ranges from \$3.27/MWhw to \$4.21 (20% wind).

(2) Balancing Cost of Utilizing ESS to Integrate Wind Power

Since the ESS capacity requirements depend on wind power patterns, different simulation results are obtained on a daily basis. The analysis results are shown in Fig. 8.8, which emphasizes the energy storage capacity requirements in 6% wind scenario and 20% wind scenario.

The results including the PCR, ECR and the balancing cost of utilizing ESS for integrating wind power are shown in Fig. 8.8. In Fig. 8.8, the simulation results from each scenario are shown in the Histogram + Probabilities graph. The "Counts" refers to the number of cases in which the requirements are in the range. The "Cumulative Counts" refers to the number of cases in which the requirements are lower than the range, and the results are represented as the probabilities. As shown in Fig. 8.8, with a wind penetration level of 6%, the ESS, PCR and ECR are in most cases less than

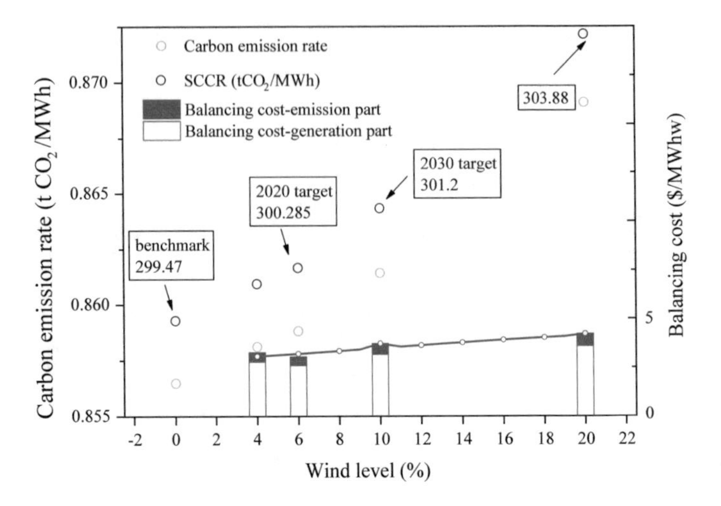

Fig. 8.7 The influence of integrating wind power on the coal consumption rate, carbon emissions and balancing cost

0.45 MW and 0.3 MWh, respectively, for a 1-MW wind power integration. When the wind penetration level increases to 20%, the PCR and ECR are 0.45 MW and 0.3 MWh, respectively.

The PCR and ECR and the corresponding balancing cost of using ESS to integrate the wind power are summarized in Fig. 8.9. As shown in Fig. 8.9, the capacity requirements are almost independent of the wind power penetration level, which is an advantage of using energy storage to integrate wind power.

(3) Balancing Cost of Utilizing ESS to Integrate Wind Power

In addition to the LCR and LCE, the costs of utilizing DSM for integrating wind power are also shown in Fig. 8.10. The LCR is expressed in the share of the standby DSM-demand in the total electric demand. The LCE refers to the expected load curtailment. The costs both in lower cost scenario and higher cost scenario are shown in the top panel of the figure. It is suggested that companies in the tertiary sector reduced demand in exchange for compensation of 1500 euro/MWh [41]. In this chapter, it assumes a load curtailment cost of $0.714 (5 yuan)/kWh (lower cost scenario) or $1.428(10 yuan)/kWh (higher cost scenario).

Analysis results show that in the 4% wind scenario, 5.54% of the total electric demand should be flexible and responsive for integrating wind power. However, the expected load curtailment energy LER is fairly small, approximately 2.69×10^{-5} MWh per MWh of wind power electricity injection. Moreover, the expected cost is far below the balancing cost of utilizing coal-fired generating units to integrate wind power. The cost advantage of DSM is obvious until the wind power penetration level exceeds 10%. However, it requires at least 5.54% of customers to participate in the DSM programs, as shown in Fig. 8.10. For 10% wind power penetration, it requires 7.92% of customers to participate in the DSM programs. Advanced metering

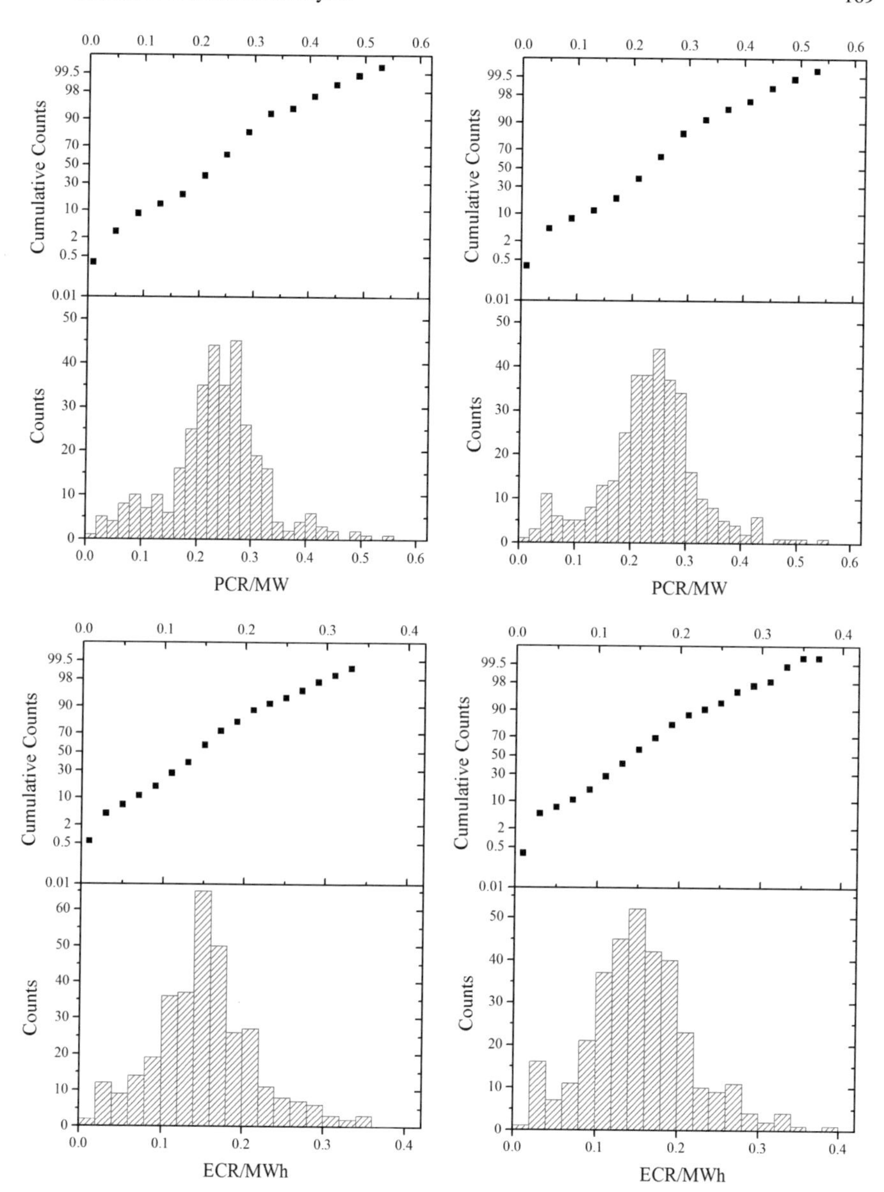

Fig. 8.8 PCR, ECR of the ESS for the wind power integration

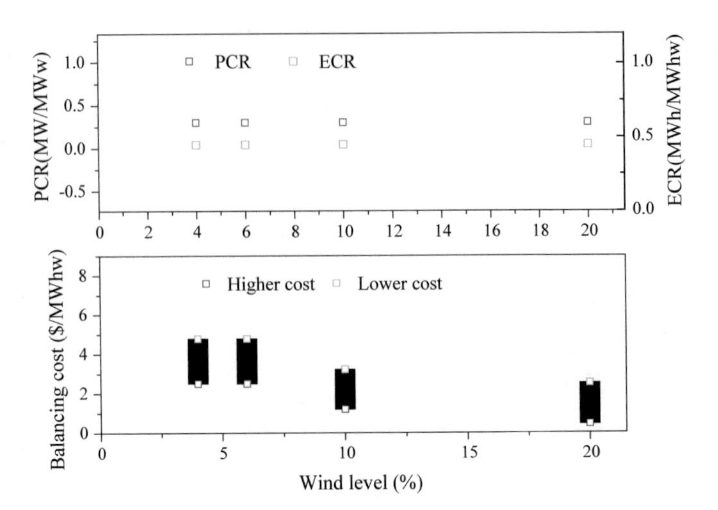

Fig. 8.9 PCR, ECR and balancing cost of an ESS for the integration of wind power

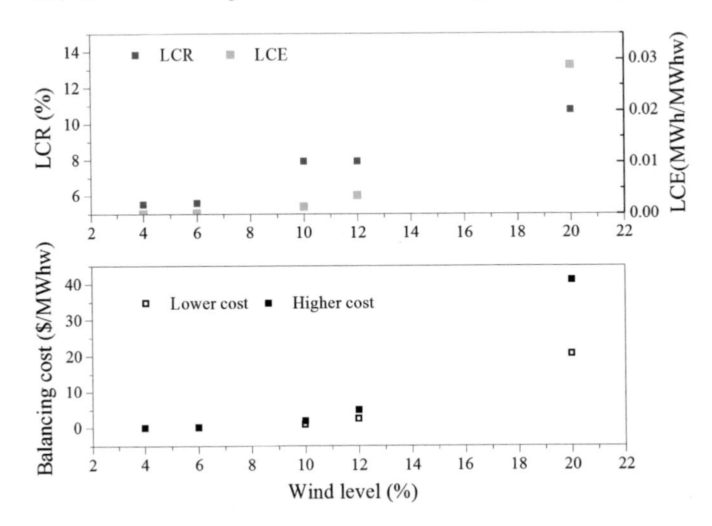

Fig. 8.10 The LCR, LCE, and balancing cost of DSM programs for wind power integration

infrastructure and the availability of dynamic pricing to customers are necessary for reaching these DSM participation levels [42]. When the penetration level increases to 10%, the balancing cost of DSM increases dramatically and makes it not a cost-effective option in higher wind power scenario.

Table 8.1 The characteristics of different flexible resources in integrating wind power

Flexible resources	Balancing cost	Depending on the wind level	Additional carbon emission	When will be cost-effective
U	\$3.27–\$4.22	Kind of	Yes	——
ESS	Depending on the cost of ESS	No	No	Cost decreases to \$400/kWh
DMS	\$0.5–\$41	Very sensitive	No	Wind level below 10%

8.3.3 Comparison and Conclusion

To facilitate comparisons, the characteristics of the primary flexible resources in integrating wind power are summarized in Table 8.1, including the balancing cost, sensitivity to the penetration level of wind power, and so on.

It is founded that the balancing cost for using coal-fired generating units ranges from \$3.27/MWhw to \$4.22/MWhw. Moreover, utilizing the conventional generating units to integrate wind power comes at the cost of additional carbon emissions. Take China for example, the increased carbon emissions due to utilizing coal-fired generating units to integrate the clean but fluctuating wind power can be 15.56 million tons per year. In other word, the effect of developing wind power is partially offset by the additional carbon emissions from the generating units providing integrating services.

Fortunately, the development of energy storage technologies and DSM offers new sources of flexibility. The increase in fuel consumption and carbon emissions can be averted by applying these energy storage and demand response technologies. Exploiting the potential of flexible customer demand is the preferred option for integrating fluctuating wind power when the penetration level is below 10%, which requires 7.92% of the customer demand to be flexible and available. Developing energy storage technologies and DSM-enabling technologies, including advanced metering infrastructure and dynamic pricing, are necessary for utilizing DSM for integrating wind power. Moreover, as introduced above, the balancing cost of utilizing DSM for integrating one-unit of wind power is very sensitive to the wind power penetration level.

Despite the current relatively high cost, the advantage of energy storage technology in avoiding incremental fuel consumption and emissions is significant. Moreover, the ESS is likely to prevail over coal-fired generating units by 2025, when the capital cost of energy storage is projected to drop to approximately \$400/kWh. Furthermore, the balancing cost of utilizing ESS for integrating one-unit of wind power is not dependent on the wind power penetration level, which makes it the best option for providing the flexibility in a power system with high wind power penetration.

8.4 Conclusions

The ever-increasing wind power production poses great difficulties in operating power systems and increases the requirement of operational flexibility. In addition to the operating reserves provided by conventional generating units, the additional flexibility requirements can be fulfilled by ESS and DSM. Considering that the deployment of the flexible resources will influence the integration of wind power technically and economically, this chapter provides a systematic evaluation of primary key flexible resources. The characteristics of the primary flexible resources in integrating wind power are founded and summarized, including their balancing costs and sensitivity to the wind-power level. The finding can provide guidance on the investment of those flexible resources to assist the wind power integration. Moreover, the methods and models are expected to serve as references for the future research in this field.

References

1. C. Feng, M. Cui, B.M. Hodge, J. Zhang, A data-driven multi-model methodology with deep feature selection for short-term wind forecasting. Appl. Energy **190**, 1245–1257 (2017)
2. H.Z. Wang, G.Q. Li, G.B. Wang, J.C. Peng, H. Jiang, Y.T. Liu, Deep learning based ensemble approach for probabilistic wind power forecasting. Appl. Energy **188**, 56–70 (2017)
3. L. Ju, Z. Tan, J. Yuan et al., A bi-level stochastic scheduling optimization model for a virtual power plant connected to a wind–photovoltaic–energy storage system considering the uncertainty and demand response. Appl. Energy **171**, 184–199 (2016)
4. M.H. Amrollahi, S.M.T. Bathaee, Techno-economic optimization of hybrid photovoltaic/wind generation together with energy storage system in a stand-alone micro-grid subjected to demand response. Appl. Energy **202**, 66–77 (2017)
5. Y. Jiang, J. Xu, Y. Sun, C. Wei, J. Wang, D. Ke et al., Day-ahead stochastic economic dispatch of wind integrated power system considering demand response of residential hybrid energy system. Appl. Energy **190**, 1126–1137 (2017)
6. G. Ren, J. Liu, J. Wan, Y. Guo, D. Yu, J. Yan, Overview of wind power intermittency: impacts, measurements, and mitigation solutions. Appl. Energy **204**, 47–65 (2017)
7. M. Kubik, P. Coker, C. Hunt, The role of conventional generation in managing variability. Energy Policy **50**, 253–261 (2012)
8. L. Hirth, F. Ueckerdt, O. Edenhofer, Integration costs revisited–An economic framework for wind and solar variability. Renew. Energy **74**, 925–939 (2015)
9. P. Simshauser, The hidden costs of wind generation in a thermal power system: what cost? Aust. Econ. Rev. **44**, 269–292 (2011)
10. H. Bludszuweit, J.A. Dominguez-Navarro, A probabilistic method for energy storage sizing based on wind power forecast uncertainty. IEEE Trans. Power Syst. **26**, 1651–1658 (2011)
11. H. Holttinen, P. Meibom, A. Orths et al., Impacts of large amounts of wind power on design and operation of power systems, results of IEA collaboration. Wind Energy **14**(2), 179–192 (2011)
12. M. Khalid, R.P. Aguilera, A.V. Savkin, V.G. Agelidis, On maximizing profit of wind-battery supported power station based on wind power and energy price forecasting. Appl. Energy **211**, 764–773 (2017)
13. C.D. Jonghe, B.F. Hobbs, R. Belmans, Optimal generation mix with short-term demand response and wind penetration. IEEE Trans. Power Syst. **27**, 830–839 (2012)

14. T. Broeer, J. Fuller, F. Tuffner, D. Chassin, N. Djilali, Modeling framework and validation of a smart grid and demand response system for wind power integration. Appl. Energy **113**, 199–207 (2014)
15. H. Falsafi, A. Zakariazadeh, S. Jadid, The role of demand response in single and multi-objective wind-thermal generation scheduling: a stochastic programming. Energy **64**, 853–867 (2014)
16. G.P. Swin, M. Godel, Estimating the impact of wind generation on balancing costs in the GB electricity markets, in *European Energy Market* (2012), pp. 1–8
17. M. Yang, R. Bewley, Integration of variable generation, cost-causation, and integration costs. Electr. J. **24**, 51–63 (2011)
18. J. Yan, F. Li, Y. Liu, C. Gu, Novel cost model for balancing wind power forecasting uncertainty. IEEE Trans. Energy Convers. **32**, 318–329 (2017)
19. M. Joos, I. Staffell, Short-term integration costs of variable renewable energy: wind curtailment and balancing in Britain and Germany. Renew. Sustain. Energy Rev. **86**, 45–65 (2018)
20. N. Mahmoudi, T.K. Saha, M. Eghbal, Demand response application by strategic wind power producers. IEEE Trans. Power Syst. **31**, 1227–1237 (2015)
21. Y. Ding, C. Shao, J. Yan, Y. Song, C. Zhang, C. Guo, Economical flexibility options for integrating fluctuating wind energy in power systems: The case of China. Appl. Energy **228**, 426–436 (2018).
22. J. Ma, V. Silva, R. Belhomme, D.S. Kirschen, Evaluating and planning flexibility in sustainable power systems. IEEE Trans. Sustain. Energy **4**, 1–11 (2013)
23. M.S. Lu, C.L. Chang, W.J. Lee, L. Wang, Combining the wind power generation system with energy storage equipment. IEEE Trans. Ind. Appl. **45**, 2109–2115 (2009)
24. P. Giorsetto, K.F. Utsurogi, Development of a new procedure for reliability modeling of wind turbine generators. IEEE Trans. Power Appar. Syst. **PAS-102**, 134–143 (1983)
25. L. Cheng, M. Liu, Y. Sun, Y. Ding, A multi-state model for wind farms considering operational outage probability. J. Mod. Power Syst. Clean Energy **1**, 177–185 (2013)
26. R. Karki, P. Hu, R. Billinton, A simplified wind power generation model for reliability evaluation. IEEE Trans. Energy Convers. **21**, 533–540 (2006)
27. H.-I. Su, A. El Gamal, Modeling and analysis of the role of energy storage for renewable integration: power balancing. IEEE Trans. Power Syst. **28**, 4109–4117 (2013)
28. N. Zhang, C. Kang, D.S. Kirschen, Q. Xia, W. Xi, J. Huang et al., Planning pumped storage capacity for wind power integration. IEEE Trans. Sustain. Energy **4**, 393–401 (2013)
29. K. Wang, K. Jiang, B. Chung, T. Ouchi, P.J. Burke, D.A. Boysen et al., Lithium-antimony-lead liquid metal battery for grid-level energy storage. Nature **514**, 348–350 (2014)
30. C. Zhao, Q. Wang, J. Wang, Y. Guan, Expected value and chance constrained stochastic unit commitment ensuring wind power utilization. IEEE Trans. Power Syst. **29**, 2696–2705 (2014)
31. T.K.A. Brekken, A. Yokochi, A.V. Jouanne, Z.Z. Yen, H.M. Hapke, D.A. Halamay, Optimal energy storage sizing and control for wind power applications. IEEE Trans. Sustain. Energy **2**, 69–77 (2011)
32. D. Elliott, Renewable energy and sustainable futures. Futures **32**, 261–274 (2000)
33. F. Liu, X. Jiang, Z. Li, Investigation on affects of generator load on coal consumption rate in fossil power plant. Power Syst. Eng. (2008)
34. C. Zhou, K. Qian, M. Allan, W. Zhou, Modeling of the cost of EV battery wear due to V2G application in power systems. IEEE Trans. Energy Convers. **26**, 1041–1050 (2011)
35. S. Han, H. Aki, S. Han, A practical battery wear model for electric vehicle charging applications, in *Power and Energy Society General Meeting* (2013), pp. 1100–1108
36. Y. Choi, H. Kim, E. Sciubba, Optimal scheduling of energy storage system for self-sustainable base station operation considering battery wear-out cost. Energies **9**, 462 (2016)
37. H. Zhong, L. Xie, Q. Xia, Coupon incentive-based demand response: theory and case study. IEEE Trans. Power Syst. **28**, 1266–1276 (2013)
38. C. Cecati, F. Ciancetta, P. Siano, A multilevel inverter for photovoltaic systems with fuzzy logic control. IEEE Trans. Ind. Electron. **57**, 4115–4125 (2010)
39. R. Doherty, M.O. Malley, A new approach to quantify reserve demand in systems with significant installed wind capacity. IEEE Trans. Power Syst. **20**, 587–595 (2005)

40. The Report on National Electricity Reliability Index in 2015 (China Electricity Councile)
41. M. Klobasa, Analysis of demand response and wind integration in Germany's electricity market. IET Renew. Power Gener. **4**, 55–63 (2010)
42. Y. Ding, S. Pineda, P. Nyeng, J. Østergaard, E.M. Larsen, Q. Wu, Real-time market concept architecture for EcoGrid EU—A prototype for european smart grids. IEEE Trans. Smart Grid **4**, 2006–2016 (2013)

Printed in the United States
By Bookmasters